"This is a lovely story of passion, persistence and patience as two refugees from pharma and finance build a business to grow plants, goats and community in Chester County, Pennsylvania. Catherine and Al Renzi delight the reader from start to finish with their challenges and stories."

— *Jane Pepper*
Former President of the
Pennsylvania Horticultural Society

"A very fun and fascinating read, Catherine and Al have channeled all of their passion, creativity and entrepreneurial experience into this book, which confidently reminds us that life is a lovely journey full of opportunity."

— *Bill Covaleski*
Founder, Victory Brewing Company

IS THE GRASS ALWAYS GREENER?

ISBN 978-1-62806-411-7 (print | paperback)

Library of Congress Control Number 2024911727

Published by Salt Water Media
29 Broad Street, Suite 104
Berlin, MD 21811
www.saltwatermedia.com

Salt Water
MEDIA

A project of Yellow Springs Farm LLC

Yellow
Springs
Farm

Cover artwork by Fuhrman Creative LLC

IS THE GRASS ALWAYS GREENER?

LIFE WITH GOATS, GARDENS, AND GOURMET CHEESE

Catherine and Al Renzi

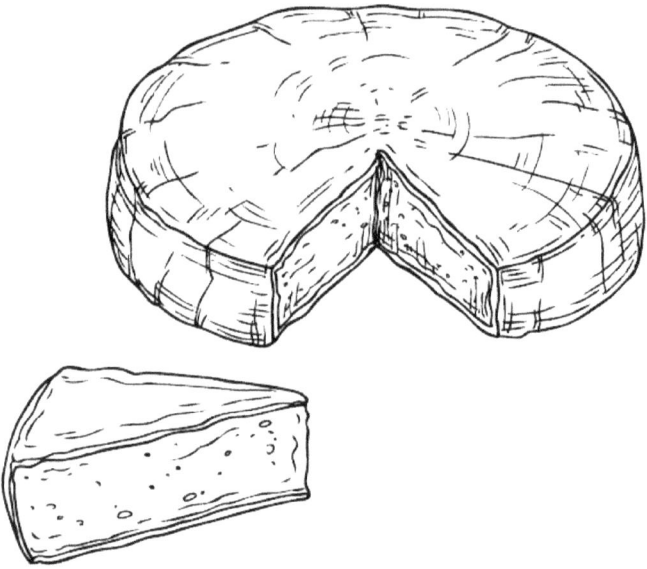

CONTENTS

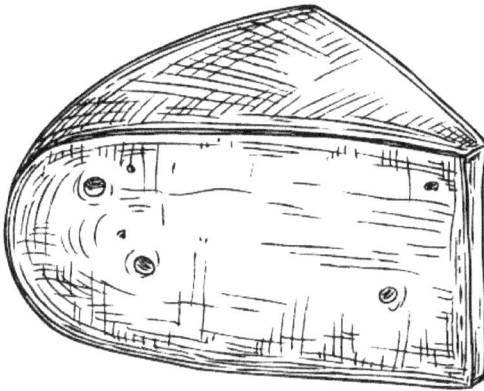

ACKNOWLEDGMENTS

We dedicate this book to our goats, now passed—Rosebud, Dora, Rena, and so many more. They live forever in our hearts and fond memories. Each goat that spent even a day at Yellow Springs Farm had a name because we believe each animal has a soul.

The farm story would not have happened without our customers, volunteers, interns, employees, cheese share members, and countless community supporters. We were better together.

Our friends, especially the members of the WACCCOs (we started out as the Wine Appreciation Club of Chester County), made sure we had good laughter, good food, and good wine when all else failed. Helping hands lifted us out of the mud, sometimes raising us as high as Cloud Nine. Read on to learn that Cloud Nine was our best-selling goat cheese year after year.

Sheryl Sensenig was our first employee. She cared for each goat as if it were the only one on the farm, and remains a dear friend. So many former employees, volunteers, interns, and customers remain close to us. They were unexpected, priceless products of the farm. We send hugs and thanks to Tess, Jen, Judy, Tricia, Bill, Barbara, Roger, Alexa, Renee, Charlotte, Divya, Christian, Lisa, Sharon, Matt, Helena and many more.

We appreciate Casey Aspin for her help from the farm's beginning to the present book project. She first connected with us as a plant nursery customer; she looked after the breeding bucks many seasons and was a cheese share member. Her singular connection to the farm story makes her editorial and technical input priceless.

Bill Cecil provided editorial expertise, and quickly became part of our new life chapter. He shared closeness with the story that only was possible because of his upbringing on a dairy farm. Andrew and Stephanie shepherded us through the logistics of publishing, and mapped our maiden voyage as book authors.

Go confidently in the direction of your dreams!
Live the life you've imagined.

- Henry David Thoreau

FOREWORD

This book is a memoir. It tells the story of a place, a space, and a time when we enriched our lives with plants, animals, and the people we met while stewarding Yellow Springs Farm in Chester Springs, Pennsylvania.

The Yellow Springs Farm property, located about halfway between Lancaster and Philadelphia, likely dates to the 1700s. When we purchased the farm in 2001, at least six decades had passed since its rolling, fertile hills supported a dairy herd. We found remnants of the old post and rail fencing rotting between the shrubs, trees and invasive vines. Occasionally, we found a horseshoe or a metal artifact and assembled them on a rack full of oddities in our kitchen. An archaeologist friend thinks that one particularly special piece, a four-inch T-shaped object with curved edges—-perhaps a handle, crank, or machine part—was sand-cast at the historic Hopewell Furnace (1771-1883), a charcoal-fired furnace about 16 miles away. The collection reminded us we were merely stewards of the farm for a period of time and that we had a responsibility to pass it on to future owners a little better than we found it.

Our research indicated a Pennsylvania German family named Himes had built the farmhouse and bank barn in the 1800s, although an earlier house might have existed on the site during the 1700s. When morning light flooded the front porch, and we looked out over dew-covered pastures, we felt connected to the families who enjoyed Yellow Springs Farm over centuries.

Not long before we "bought the farm," we had exchanged marital vows in an Episcopal church in 2000. The formality and finery of the celebration stood in contrast to the mundanity of a real estate transaction; however, the vows on both occasions were equally profound for us. We did not have children. The farm and all that happened there from 2001 to '21 inadvertently became the mission and product of our marriage. We shared a home, a dream, and a business.

As partners in marriage and business we finished one another's sentences, predicted one another's needs, and knew that when all else failed, we always had one another's back. This was why and how Yellow Springs Farm persevered for 21 years. The farm continued because, despite the difficulties, we never both quit on the same day.

As you read this book, please understand that we use the first-person plural pronoun "we" throughout to capture both of our experiences. Like everything else on our farm, we worked together to create each chapter, mixing our writing and memories. When clarity merits specific personal information, Al or Catherine is named as a second person, perhaps sitting across the room. The space between Al and Catherine Renzi is hardly there. It is a composite of overlapping actions, decisions, and emotions. Al and Catherine Renzi lived at Yellow Springs Farm as a compound noun. It is impossible to separate one from the other without materially changing reality, so we encourage you to embrace Yellow Springs Farm as we know it together. We always had one business card with both our names on it. This reflected the complete faith and confidence we shared to be one for all, and all for one—always.

EASTERN PENNSYLVAN

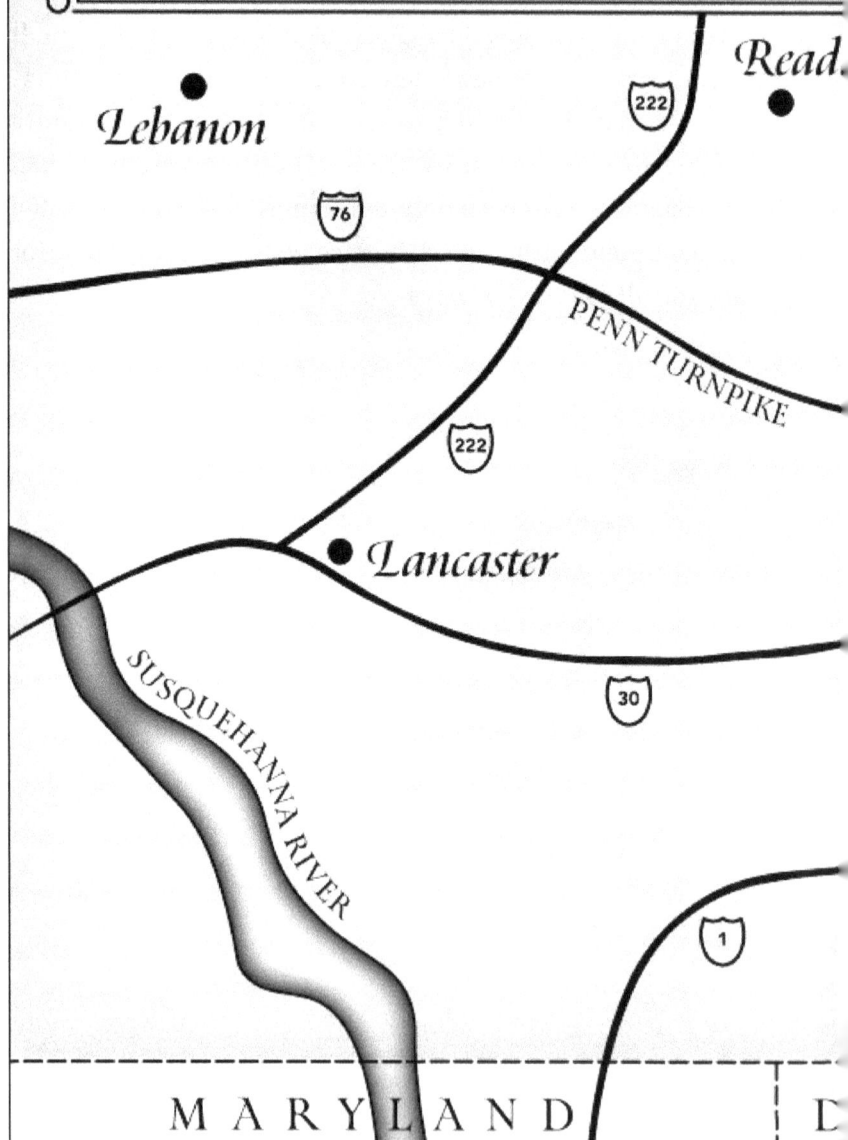

Read.

Lebanon

222

76

PENN TURNPIKE

222

Lancaster

30

SUSQUEHANNA RIVER

1

MARYLAND

D

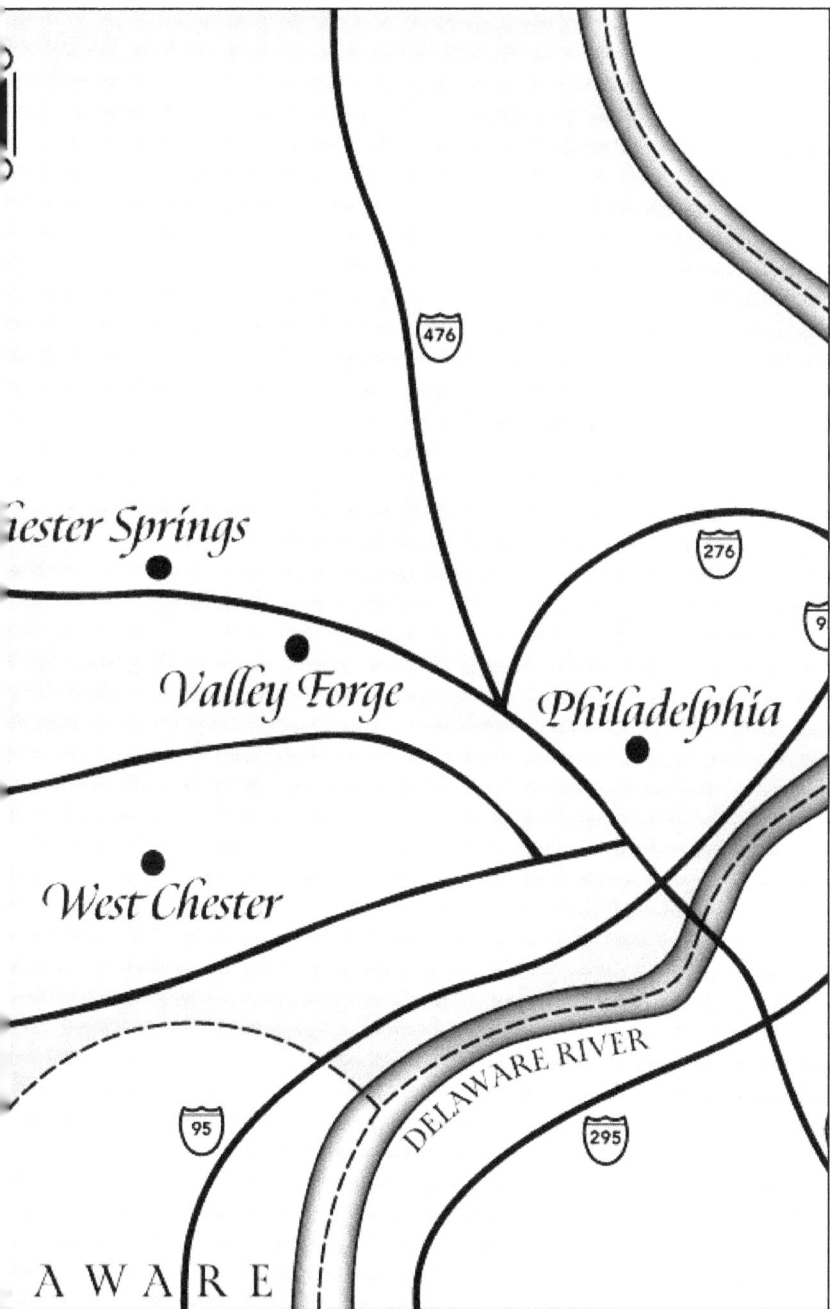

CHAPTER 1:

IF NOT NOW, WHEN?

Why do two middle-aged career changers with five college degrees between them buy a farm? Most people who don't know us well assumed we sought a humorous respite from harried lives rushing to keep pace in corporate cubicles, professional conferences, and cyberspace. Al worked in business development roles for pharmaceutical and biotech companies. Travel was his norm. His starched shirts and dark suits were perfect, but somehow losing their luster. Catherine spent her days on the phone and in meetings, working in financial services. Her favorite monthly appointment was meeting the blacksmith at 6 am for her horse to get new shoes before the business day began.

Corporate life was challenging for both of us. It was not only the business travel, but the need to fit into a box with particular expectations. Navigating the performance

metrics, organizational dynamics, and politics of office culture was not something that came easily for us. We were not
doing ourselves a favor individually or as a couple by staying in jobs that, although intellectually interesting, were not
emotionally fulfilling. Nevertheless, we were responsible,
practical adults. We were not planning to purchase an 1850s
farm. There are no courses to take or how-to books that prepared us for building a native plant nursery, honey-produc-
ing beehives, an organic vineyard, and a goat dairy making
artisanal farmstead cheeses.

Before moving to Chester Springs, Al lived in Kennett
Square, Pennsylvania, where suburbia met the countryside.
Strip centers marked some intersections, while others featured historic barns, springhouses, or stone walls. His running route after work was punctuated by pastures with split-
rail fences. Al saw these landscape features as property of
the 19th century residents, the Brandywine River Museum,
and Andrew Wyeth's (1917-2009) imaginative, vernacular
paintings depicting a narrative of life in Chadds Ford, PA,
and surrounds. Al did not quite understand how best to in-
tegrate his corporate commuter lifestyle within these bucolic settings, but he knew that there was a place for him.
What he did know was that he loved the idea of living in an
old Chester County farmhouse surrounded by open space, a
chance to breathe fresh air and to walk on the land traversed
by the footprints of history.

In contrast, Catherine dreamed about owning a farm,
but felt it would be something that would have to wait for
decades. Perhaps it would be a retirement project? In the
meantime, she resigned herself to her backyard garden, two

cats, and postponing self-actualization. The itch for pets and open space had taken root during her younger years, when she explored the countryside on horseback in Pennsylvania, New York, Vermont, Massachusetts, and Virginia. The study of art and architecture, plus time spent living in Italy, caused Catherine to gravitate towards historic stone buildings. Like Al, Catherine saw these picturesque scenes in her life up to that point as belonging to others but had not considered that she might soon live in the landscapes of her mind.

While dating in 1999, we heard about a photo contest in Pennsbury Township. This was close to Al's townhouse, and an area we both knew well. The annual hot-air balloon festival in late summer would feature the displayed entries. This impetus prompted us to drive slowly on miles of Chester County rural roadways. We pulled over to photograph animals, barns, farmhouses, and covered bridges. More than once, we snuck in a kiss. We fit in nice picnic lunches, listened to music on the car stereo, and both shared the camera as we captured the scene. Weekends passed quickly. We grew closer and were falling in love. After printing some photos in color and others in sepia tones, we selected frames and surprised ourselves with how the work we did together was clearly better than what either of us would have accomplished alone. The photo contest was our first shared creative project. We did not win a prize, but it set the tone for joint ventures to come, including marriage, Yellow Spring Farm, writing this memoir, and much more. Separately, neither Catherine nor Al would have ever even come close to buying the Farm; together, everything was possible.

Our love story was romantic, but not without aspiration

and ambition. When we married in 2000, Al sold his house, and we lived together in Catherine's suburban Cape Cod in Newtown Square, PA. It was about a 40 minute drive from Yellow Springs Farm. We were looking for a house together so we could begin our new chapter on fresh ground. We spent months of weekends looking at houses in PA in summer, fall and winter, but did not find the right place. Tired of the real estate routine, we celebrated our first wedding anniversary in St. Michaels, Maryland with a romantic three-day weekend over Memorial Day 2001. This choice of a Maryland getaway location reconnected us with our first "away date" on the Chesapeake only eighteen months earlier. Driving back to PA, we detoured to drive by a Chester County property for sale that we had visited with our realtor twice in recent months, but rejected. Since then, the price was dropping, and the spring weather made it look much more attractive than it did when it was covered with snow and ice. There were hundreds of reasons not to buy the Chester Springs, PA property, but the enchantment of what might go right overcame rational calculations and risk adversity. We were confident the risks were worthwhile, and the adrenaline produced during the journey forward would be the cure for occasional doubt, nausea, or anxiety. Dreams and romance danced in our imagination in July 2001 after we closed on the farm purchase. It came complete with a mortgage and a 26-page property inspection report that served as our first to-do list. When we admired the daylight from every window in the farmhouse, walked in the woodland, and sat around the swimming pool nestled behind the barn, we almost expected a stranger would appear and ask us to leave. "Could this really be our home?" we wondered. Looking back from the distance of decades, the decision still made little sense, but we have no regrets. We traded manageable predictability for an idle farm with three stone buildings in need of restoration and repairs.

Al had only had a fish tank and a turtle before dating Catherine. Al got a hint at what farm life was going to be like when Luigi the cat was licking his eyelids at four in the morning hoping to be fed early, and when Snowball—our outside cat—climbed a tree in the middle of the night, banged on the second-floor bedroom window and demanded to come inside to the warm house. In Catherine's company, Al quickly learned to carry a lint roller for shedding season, buy muck boots at Tractor Supply or LL Bean, and enjoy the chaos of furry family members. Catherine boarded her horse about 30 minutes from the house. Most of our dates during a succinct year of courtship happened before, during, or after time spent at the horse barn, or cleaning the litter box. It is not clear if Al immediately loved his quick introduction to animal care and barn chores while dating, but we definitely loved one another.

When we tell stories about our courtship, Catherine reveals she knew Al would be the one during the many animal encounters – the way he cared about the cats and horses, and invested time and attention to get intimately acquainted with each furry friend's preferences, personality, and hot buttons. Al undoubtedly was smitten once and for all when he and Catherine were out for a casual dinner after a session at the horse barn. She asked as they were leaving, if he wanted to see her new (but pre-owned) truck. "Sure," he said, as they stopped in the parking lot near the tan 1997 Ford Expedition Catherine just acquired, so she could independently pull a horse two-horse trailer and travel to horse shows. Catherine proudly exclaimed, "It even has a Triton engine." "Perhaps the best engine Ford has ever made," she added. Al did not know anything about trailer towing vehicles or Triton engines, but his heart melted over the only woman he ever met who could hardly wait to show him her truck. The engine specs were icing on the cake.

We fell in love at first sight with Yellow Springs Farm in Chester Springs, but learning to know this place was a slower, more nuanced process of looking, seeing, and learning. We figured we would take things one year at a time, and eventually find more momentum restoring the property's agricultural heritage closer to our retirement years. In December 2001, months after the purchase, we donated a conservation easement on the property to the French and Pickering Creeks Conservation Trust. Eleanor and Sam Morris founded this local non-profit organization in 1967. Its work has preserved over 13,000 acres of open space in northern Chester County, Pennsylvania. A conservation easement is a legally binding document that protects the farm pasture, woodlands, pond, waterways, and farm footprint from subdivision and land development in perpetuity. Essentially, the landowners agree to donate land development rights to a land trust or municipality in exchange for tax deductions. Easements attach restrictions to a property such as a defined building envelope, prohibitions on removing trees, and tight limits on coverage by impervious surfaces.

While dating Al, Catherine used to cut out articles about land use topics and other shared interests. One day, Al received in the mail a soon-to-be fortuitous article from the Wall Street Journal that outlined the financial planning aspects of conservation easements. When we married in 2000, we both aspired to find some way to be stewards of the natural environment and preserve a farm or landscape. Purchasing and protecting the farm with a conservation easement realized this aspiration for us.

During the process of completing the easement, a botanical survey of the property identified that one-third of the plant species on the eight-acre farm were invasive plants. We were bombarded by this knowledge, but unsure of the next steps. What could we do to improve botanical biodiversity? Were we doomed to live with the invasive species? We talked to local horticultural experts, read some books, and attended symposia at nearby arboretums and nature centers. We visited parks and preserves where native plant communities were better established, and began to understand what might be possible to achieve in the gardens and naturalized areas at our farm. This inspiration prompted us to start a native plant nursery. It fulfilled a desire to start our own business, utilizing our educational backgrounds, and would eventually bring the farm back to agricultural use. Creating a native plant nursery seemed within reach, in contrast to the intimidating capital investment needed to restore the farm's historic dairy heritage.

Al and Catherine were both accomplished gardeners even before they owned Yellow Springs Farm. Both grew up in families that faithfully kept kitchen gardens for staples such as tomatoes, cucumbers, and culinary herbs. Al took continuing education courses at Longwood Gardens, an internationally-acclaimed botanical garden and arboretum in Kennett Square, PA. Al had a small bonsai collection, and was an expert at container gardening, too. It showed in his colorful urns and decorative clay pots filled with blooms on the deck and patio. Catherine studied perennials and aspired to garden every inch of her small backyard before the farm purchase. She built small stone walls, designed new beds, and found weeding to be a meditative, calming pastime. Gardening was a common hobby both shared before marrying one another, but farming would become the start of whole new life chapter.

No one needs a reminder about what happened on September II, 2001. Al and I left our offices early that day and convened on rocking chairs under the farmhouse's covered front porch. There were no goats and no fences, but plenty of long pasture grass in the vista. We heard and saw only occasional military aircraft overhead, but looked up anytime there was a sound. We knew what had happened in Manhattan that morning, and were only beginning to understand details about the plane that went down at the Pentagon, and Flight 93's end in Shanksville, Pennsylvania. In these hours, nowhere seemed completely safe, but the farm offered the only respite we could find from frightening world events.

Mourning for the loss in our community and nation characterized the next several weeks. There was an underlying traumatic fear that the tragedy might not be over. What acts of terrorism would happen next, and where? When the U.S. airspace opened, Al boarded one of the first flights from Philadelphia to London, as he had been summoned to a business meeting. He called Catherine from the airport and told her he had been through several extra security screenings, that the plane was almost empty, and that air marshals were among the few passengers. Catherine stayed awake all night to monitor the plane's route on radar as it crossed the Atlantic. She felt relieved when she saw that it had landed safely. Corporate jobs offered steady paychecks, but often required rigid schedules, and sacrifices of personal time with frequent business travel. The ordeal begged the question: "Is it worth it?"

We had been married just over one year. We had so much to live for and feared there might be little time to squander. Current events in 2001 made "carpe diem" more than a Latin phrase excerpted from an ancient Roman poet. It became an urgent mantra to seize the day, start now, and forge ahead. We started spreadsheets with a plan for every contingency and we revised the calculations every month. After three years on the farm, we learned that spreadsheets controlled nothing—life happened without rhyme or reason. With more courage than convincing strategy, Catherine left her office job in 2004 and began spending all her time on the farm. "Leap and the net will appear," she thought, and so it did then and again when Al came to the farm full time in 2006.

For several years following the events of 9/11, while spending time in airport business lounges, corporate center parking lots, and pharmaceutical trade shows, we took on property renovations by the dozen. Al's work exposed him to armies of executives driving fleets of expensive cars—Jaguars, Mercedes and the like. While he appreciated the aesthetics of these sleek machines, he stuck to a modest Honda, allowing us to plow his company car allowance into the farm, everything from historically accurate barn door hardware to masonry restoration work. When it came time for us to buy our first vehicle together, we were similarly practical: enter one orange Kubota farm tractor.

As Al drove the tractor around the farm, he was unknowingly driving farther and farther away from the pharmaceutical industry. He thought the farm could become a place to use his science and business experience productively, but it was unclear what that would look like. Many of these musings, daydreams, and daring hopes were silently processed, and shared only as tidbits of passing thoughts sprinkled into conversations over many months. Al and Catherine were often apart during the first years at the farm. Both travelled for business, and on frequent occasions returned home late. When we stole hours together, there were household plans to make, budgets to review, and difficult physical outdoor tasks, such as clearing vines, splitting firewood, and repairing gates, necessary to complete before sunset.

Our workload was soon to become a little lighter, with power equipment on the way. The Kubota was brand new when it was delivered to the farm. The tires shined, and it seemed neglectful to drive the tractor across the gravel driveway in case a stone spun up and nicked the paint. Good thing that we planned and bought a couple of cans of Kubota-licensed spray paint at the dealer's shop. We felt prepared to take care of our new baby, but quickly learned a few things about having our priorities straight when the oil needed to be changed, and shear bolts needed to be replaced. If it were only as simple as selecting the paint!

Catherine looked around at tack shops that sell name plates for horse stalls, but could not find a nameplate sized to fit the Kubota. Still wanting to name and baptize Al's "new car," Catherine visited auto accessory stores that apparently had never had a female customer, or so it seemed to her after having the misfortune of using the store rest room. We gave up the search for a bronze plate, and continued referring to the tractor as "Big Orange."

Workshops for new farmers sometimes have sessions on how to repair a tractor, but we chose instead to opt for a good warranty, given the limits on our time and resources. To keep the house and barn from falling apart, we focused on learning a few carpentry, plumbing, and electrical skills. With 36-month interest-free financing, we hoped and prayed the Kubota would be virtually maintenance free, at least until it was paid for.

We increasingly came to depend on the Kubota. We used it to move stones, plow the driveway, mow the fields, and dig holes for tree planting. So valuable was the machine that we added a "Big Orange" rule: Every year after we turn 40, we purchase a new attachment for the Kubota. The goal was for the Kubota to assume the physical tasks that our aging bodies protested against. Circumstances taught us a few more tractor basics, but not exactly mechanics. First, keep a tire patch on hand always. Second, keep a compressor on hand to fill the tires to make the tire patch more useful, especially since the tractor is not titled for road use and the nearest gas station is a few miles away. Third, don't forget to put fuel stabilizer in the tractor (and the chain saw, the mower, the trimmer, the shredder, and all the fuel-hungry farm tools.) These details saved some time and money, but our most important lesson was to keep the hood raised on the Kubota during the fall and winter. We learned this lesson from Quint, a Doberman/German Shepherd mix rescued from the SPCA. He circled the Kubota, barking and scratching at it; good thing the Kubota orange spray paint was handy. Dense, but not clueless, we dumb humans finally lifted the hood to find a mouse nest among the engine, hydraulics, wiring, and other very essential parts of the tractor the farm had become so dependent on. We learned the Kubota warranty did not cover rodent infestation.

Quint got an extra treat and profuse verbal praise for that call. Unfortunately, he couldn't help us with the utility trailer, which was parked for the winter up the back driveway, well beyond the range of the dog's invisible fence. When the trailer went for state inspection in early spring, we found our Kubota mice had relocated in the rear lights and wiring of the trailer. A bumper pull trailer is a relatively simple piece of farm equipment as these things go. It has no engine and no hydraulics. Nevertheless, the trailer repair bill was several hundred dollars. How frustrating! You can't sue a mouse for damaging your property, nor petition a local government to enact an ordinance to prevent squatting mice within the borders. In short, the dog was more than earning his keep, and the farm owners were learning the hard way.

Friends and former colleagues were concerned for us

as we told them about our transition from corporate life to farm life. They worried about how we would pay the bills, and if our marriage would survive being together almost all the time. They asked if we missed our former lives, or if there were any regrets. Eight years into farm life, they were still asking, but much less often. There were nervous congratulations from a few folks who admired our temerity, and some muttered lines that sounded a lot like condolences from those who feared this lifestyle change was a path peppered with danger and likely ending in a near-death experience.

Al's analytical, methodical mind allowed him to appreciate the farm as if he were looking down from 30,000 feet. He is capable of the sort of detachment that Catherine's passionate and emotional spirit doesn't experience in quite the same way. Al mused, "I am a believer in coming full circle in life. I believe we start at a place in our life for a reason and then end up coming back to a variation of that same place, in a relative sense, at least." Our choice of a more rural lifestyle, similar in some ways to what our great-grandparents knew in Italy, was perhaps a nod to our Italian-American immigrant families. We both grew up in families that valued education as a path to socioeconomic upward mobility. This version of the American dream lauded hard work, but, above all, prioritized raising children to become first-generation college graduates.

Perhaps it's DNA, or maybe lessons learned by growing up in entrepreneurial families. Catherine's grandparents had a luncheonette, a pub, and a wholesale produce business. Al's father was a tailor by trade, and this eventually evolved into Al's parents purchasing a clothing store where they worked together for many years. Al saw first-hand how his parents worked together. They were role models for how we would become not only partners in marriage, but also business partners.

One of the farm tasks that Al took ownership of was going out with the truck and trailer to buy hay and straw from Ray at a nearby farm. Al loved telling stories about Ray's place. The guys would sit on straw bales in the loft and talk about the weather, the rain or lack thereof, and the latest commodity prices trends. Ray raises longhorn cattle for beef and is as passionate about them as we are about our two Nubian goats. Imagine sitting in a hay loft around sunset and looking down on a steer with six-foot wide horns grazing while the scent of fresh-cut grass fills the air. Al truly enjoyed going to get hay. It certainly beat sitting around a conference table in a boardroom.

But corporate life is poor preparation for farming, as we had already discovered with our spreadsheets. A minor example is Al's limited hay knowledge. When Ray wasn't around, Al needed to know his hay varieties: alfalfa, orchard grass, or timothy? Timothy hay has the "light-colored fuzzy brushes" at the ends of the stems, Catherine counseled by phone. Horses like that, but goats prefer alfalfa. Al looked for greener, finer textures in some bales as a barn cat leapt and sauntered among the stacked rows, enjoying the game of hopscotch. Al thought he might have it right, so he called again, and was told that if it were too fine and soft, that was grass hay. Alfalfa would be greenish, but stiffer and coarser than the grass hay, but without the fuzzy ends of timothy hay. Catherine asked, "Why don't you smell the bales and decide that way?" In hindsight, that was akin to asking a colorblind person to choose the shirt that looks best with the pants. Thanks to the good graces of unlimited calling minutes between two phones on one account, we had enough time to discuss detailed descriptions back and forth multiple times. This ensured that we loaded the correct hay, and the goats would be happy and well-nourished.

Later that week, Al and I had one of our periodic, in fact fairly frequent, gut checks. Without being planned, just out of the blue usually, one spouse asked the other, "Are you doing okay? Are you having fun? Having a good day?" For the first several months after Al left his corporate role, Al's reply to our marital check-ins often included a reference to Ray's farm. Al was taken aback by the work, the accomplishments, and the choices he observed in Ray's farm routine. Al and Ray were in the same age bracket and lived just a few miles apart. Ray was a third-generation farmer; his brother farmed, his father farmed, too. He owned land, farmed leased land, raised field crops and livestock. The contrasts in upbringing, backgrounds, and aspirations were many, but Al connected with the feeling that farm life was the right life now.

CHAPTER 2:

HUMBLED BY THE HOOPHOUSE

We repeatedly reviewed the botanical inventory completed for the farm as part of the conservation easement process. Learning to identify and differentiate among the native grasses and ferns was challenging. Catherine has an artist's observant eyes, but still struggled with counting stalks, leaves, and plants' reproductive structures. How could we remember which species had alternate, not opposite, buds? Wildflowers' sequential blooming, even with each species in flower for only a few weeks per year, was a textbook open right before our eyes. Trees with leaves were straightforward to identify, but naming tree species by only the winter bark was harder to learn. Redbuds were undoubtedly our favorite plant species in all seasons of the year. The spring flowers were magnificent, the yellow fall color accents the landscape, and the parasol formed by bare branches arching over the snow was equally stunning. We had Redbud trees in the front woodland, and the rear hillside, but started adding more to the roadside, patio garden, and hedgerows. Was it just a coincidence that the first Nubian goat purchased a year or so later was already named Rosebud? The alliteration is at least synchronicity, or perhaps a hint that everything was going to fit together well.

Our first winter on the farm was like being in an enchanted, idyllic scene. Our dogs, Quint, and Rugby, a Rottweiler/Swiss Mountain Dog, would join us by the wood stove as we watched the winter sky from a large window in the great room. Sunny weather enticed us to explore the woods or take a leisurely stroll by the pond. The dogs established a "no-fly zone" in the farm's airspace. They announced every hawk, turkey vulture, and heron without fail. Only the herons would alight, as the pond provided a dog-free zone. After being on "cloud nine" for several months, we realized we had to do something other than wander around the property admiring nature. If it were to become a native plant nursery, research and hard work had to start soon. Unlike the change of seasons, this would not happen on its own.

It made sense to visit a few established retail and wholesale native plant nurseries. The first cold calls were slow, but most people were surprisingly open and enthusiastic about having us visit. We quickly realized the market research we were undertaking was an enormous improvement on the sterile statistics and stuffy meeting rooms we were accustomed to. We felt accepted and encouraged by the nursery owners. These were people who shared our values, hopes and dreams and they were enthusiastic to help us get started.

To start, we visited a wholesale native plant nursery in Lancaster County, Pennsylvania. Jim's wholesaling business had thrived for over 10 years, largely by focusing on restoration projects. Wholesalers sell to landscapers, government entities, and other institutions, but also supply plants to retail nurseries. Especially since 2000, there has been a large movement by the public and private sectors to restore municipal parks, open land, stream banks, and roadsides with native plant material. In some localities, local ordinance mandates the choice of native plants. The publication of "Bringing Nature Home" by Doug Tallamy in 2009 marked a milestone in bringing public attention to native plant species, and that momentum continued today.

The wholesale nursery has acres of land with thousands of plastic pots, many hoophouses, and extensive irrigation systems. The staff included local, plain sect women, Mennonites and Amish, wearing traditional bonnets. It was a pretty amazing sight, but also very humbling. Yellow Springs Farm had eight acres of property, and much of that were wetlands and woodlands. We could never have enough land to run a business like this, and we were not even sure that was what we wanted. Jim was encouraging; he offered us his shelf of catalogs to research suppliers of hoophouses, pots, and soil. We all agreed there was an opportunity in retail plant sales, given there were few native plant nurseries in southeastern Pennsylvania and nearby states. We knew this from our own experience, too. When we looked for native plants to restore the farm woodland, there was not a nursery within a 45-minute drive to help us, and even those an hour or more away had only a small portion of the plants we needed to restore the landscape.

Yellow Springs Farm native plant nursery was beginning to take shape. It would not be the typical retail shop. Most plant nurseries have regular hours and are open seven days a week, at least during the gardening season. When we visited local retail nurseries, we realized it was not the type of lifestyle we had in mind if we were to exchange our business suits for denims. High overhead costs often burden retail operations. The difficulty of finding and keeping low-paid, part-time help, and the wasted hours while waiting for a customer to come to the gate add to the problems of retail.

We liked the ethos of the sign we saw at a business in a nearby village: *"Open most days about 9 or 10. Occasionally, as early as 7, and sometimes as late as 12 or 1. We close about 5:30 or 6. Occasionally, about 4 or 5, and sometimes as late as midnight, or later. On some days, we're not here at all, but lately we've been here a lot, unless we're not here."*

Next, we visited a retail nursery near the Pennsylvania-Maryland border specializing in native plants. The owners were friendly enough, but our unscheduled visit corresponded with one of their open house dates. We were pleased for their sake, and hopeful for our own future, because the event drew a crowd. The nursery staff was understandably too busy to show us around. We spent a little time talking to them and picked out plants to purchase. The red waving wands of Lobelia cardinalis, and the floriferous asters in many shades and sizes were some of the irresistible native flowers. Overflowing black plastic trays were loaded in the SUV, so the plants far outnumbered the human passengers on the drive home. The visit helped us see that a farm-based nursery could work. It fit our lifestyle. We would not have to be open all the time, and the farm itself was a marketing tool. Perhaps having a 1850s Pennsylvania fieldstone farmhouse and bank barn, pond, springhouse, and menagerie of furry pets might give us an edge. If only it were that easy!

With a business model selected, we were eager to grow some plants. We met in the Yellow Springs Farm conference room, AKA our dining room table. As we discussed plant selection for our first plantings, a black streak flew past the window, and then an equally fast red streak. We let our imaginations run wild for a spin, but then realized a red fox was in hot pursuit of our barn cat, Chloe, as the cat and fox raced towards the pond. Its surface looked like solid ice. Chloe was smart enough and light enough to traverse the ice and flee. The fox was not so lucky. It fell through the ice for an undignified swim. Another day at the office!

The summer of 2002 was going by too swiftly. We were both still working long hours at jobs off the farm and planning the nursery start-up activities, mostly on weekends. The capital costs were theoretically less than what we eventually learned about starting a dairy or cheesemaking business, but it certainly didn't seem that the nursery start-up was cheap or easy. First, did we need a greenhouse—an expensive investment in glass and foundations? We decided on a hoophouse, essentially a greenhouse without heat. It was cheaper and consumed no energy or fossil fuel, so it was the sustainable farm's perfect fit. It had no glass, just a bunch of steel hoops we had to assemble, anchor in the ground, and cover with a mesh shade cloth in summer and plastic in winter. This seemed simple enough.

We considered hiring help to assist with erecting the first hoophouse at the farm. Isn't it almost always wise to learn a new task by watching an expert do it first? Since we couldn't find any relevant contractors, we became convinced that humans likely possess a genetic trait inherent in them to put up hoophouses. How hard could it be?

For a husband-and-wife team with no prior experience, it was quite difficult. A hoophouse is 20-feet wide by 50-feet long and includes no interior walls or utilities. Our hoophouse was shipped dissembled in a carton box from Canada. It arrived with instructions in French and English—but mostly French—a bunch of hoops, screws, and end posts, plus many other trappings we could not name.. We learned the hard way that calling the customer help line was bad news. Turns out a call to Canada on a number that was not toll-free was very expensive, particularly galling, given that most of the time we were on hold. After almost half a day's work, we could not agree on how to site the corner posts. Catherine called on her experience on horse farms setting up dressage arenas. She relied on the Pythagorean Theorem—A squared plus B squared, the two sides of the corner, equals C squared—the length of the diagonal between the end points of the corner. Right angles in each of the four corners are critical in dressage arenas for symmetry so that competitive riders can practice balletic equestrian performances. The sequence changes occur at precise points according to the letters marking the perimeter of 20x60-meter rectangles. It is perhaps best compared to a compulsory figure skating routine.

We both agreed that if we didn't get the corners right,

the structure would not be square, and the hoophouse would fall. To sell her method, Catherine wandered off and created a right triangle out of some scrap wood found nearby. She measured, cut, and stapled the pieces together. After more doubts, and some disbelief on Al's part, the corners did square up and the project was on its way. Over a series of weekends, we brought in stone, leveled the ground, and almost completed the nursery.

After toasting our new pseudo-greenhouse and admiring our "yes we can" attitude, we found out we needed fans for the hoophouse to keep the air circulating. After all, plants are alive and breathing. Basic biology predicts that if you put plants in a small space with a cover on it during humid summer days or even damp fall days, the plants suffer. Fungus, mildew, insect pests, and diseases thrive, but plants don't grow. One minor mistake was not a problem, but part of the normal learning curve, we reasoned. We bought a couple of circulating fans and mounted them in the hoophouse. Mission accomplished, right? Not quite. We quickly realized the importance of having an exhaust fan to eliminate the circulating air in the hoophouse and bring in fresh air. As aspiring horticulturists, we should probably have known that circulating stale, humid air filled with mold spores and soil fungus was not the way forward. Like the voice-over in a late-night TV commercial said: "That's not all, folks!" We needed to install an electric sub-panel, design an irrigation system, trench water lines from the well, and select landscape fabrics. We just wanted to grow a few plants! By word of mouth, we tracked down a contractor named Baltha who had put up a hoophouse once before. He added the necessary finishing touches. Thousands of dollars, countless hours, and numerous specialty tradespeople later, the nursery was set to go. It was time to pot-up our first crop of seedlings in early fall so they would be ready for spring sales.

Waiting for the fun part to start, we headed off to a nursery that sells starter plant material. These are typically 2-inch plant plugs sold in 50-or 72-cell trays. If we did it all from seed, we would never have enough plants to sell in the spring. This was the only way we knew to get things going quickly. We met with the sales rep, studied plant lists, and made a best guess about what would sell to our customers-to-be. Many of the first species we selected remain among our favorites today: False Indigo, Brown-eyed Susans, assorted ferns, and Purple Coneflowers. The black plastic pots were neatly arranged in the hoophouse according to size; they resembled the Three Bears—Papa Bear was the 1-gallon size pots, Mamma Bear was the quart pots, and Baby Bear was the smallest seedling. Perhaps it was at that point that the Yellow Springs Farm business was real.

We puttered and labored over the pots, watering them carefully, and tending to their every need even though we were not always sure what their needs were. As Thanksgiving approaches in our region (then officially USDA Zone 6), tender plants in containers need to be protected, or they will die from the cold temperatures. More precisely, the freeze and thaw cycle that comes with wildly oscillating temperatures damages even hardy species. Global warming might someday soon make it less cold in Zone 6, but it is unlikely that the weather will become more predictable.

Two days after Thanksgiving, we dutifully covered the

hoophouse with an ugly white plastic. Everyone assured us this was a proven practice and that the plastic would be re-usable in future years, and eventually recyclable. It was an important sales pitch for a business built around sustainability. Based on experience with failed swimming pool covers rated with misleading names such as "Five-Year Solution" and "Seven-Year Guard" it seemed prudent to focus on the recyclable aspects of the plastic's testimonials.

The daylight hours grew shorter, and the holiday season arrived. The native plants were dormant in their pots. All was going according to plan. It was a relief to think, at least for the first season, that there is not much to do at a plant nursery during the winter except wait patiently for spring. A variety of farm building renovations were planned before the farm business launch the following spring. We created a website using the Yahoo e-commerce site and printed a brochure with help from Ellie at the Gecko Group. Plant tags with pretty color photos and Latin names we tried to pronounce correctly were crucial to our future retail marketing strategy. Thinking of future sales days, we were ready with Yellow Springs Farm t-shirts, a market canopy for shade, chairs, receipt books, a cash box, and even business cards.

Nearby West Chester had a growers market, a weekly market where stall holders sold meat, produce, prepared foods, and some garden plants. We attended the winter organizational meeting of the market and were delighted to learn that we could join the vendors in the spring of 2003. We also planned a two-day open house at the farm for later in May, on Memorial Day weekend. Significantly, we would celebrate the life we created together on the weekend of our third wedding anniversary.

The first market day in May was to be a sort of coming out party for our fledging business. We checked boxes on the business plan, reviewed the marketing plan, and documented each step as our corporate training taught us to do. March rolled around and, as temperatures rose and days lengthened, we could see the quiet stirring of the plants as they formed new growth at the base of the crown. It was an exciting experience for us. We were so afraid we might have killed all the plants with kindness, neglect, ignorance, or some combination of the three. We were reminded again that Mother Nature does not require us humans to teach her how to keep things alive.

Market opening day was here, and we packed the horse trailer with supplies Friday evening after work and loaded the plants at the crack of dawn on Saturday morning. Despite having logged a week at our desks, joining the market energized us. We were both excited and petrified at the same time. What if no one bought the plants we worked so hard to grow? We set up displays and signs and primped the plants as if they were poodles about to enter the ring at the Westminster Kennel Club show. Lo and behold, the bell rang promptly at nine o'clock, signaling the market was open for business. Since the growers market had never showcased native perennial plants before, people initially saw us as a curiosity. We chatted, smiled, and worried a little. It seemed like an eternity or an instant depending on when you asked, but within fifteen minutes we had our first sale, and then another, and another. When the closing bell rang at 1 p.m., we felt a surge of adrenaline as our sales surpassed our expectations and eased our worst fears.

We had cash and some checks in the box, but our first market day gave us much more to be grateful for. We created new connections with the customers who shared our values. At the time, we did not fully appreciate that customers attended the West Chester Growers Market not only to buy but also to support their local community, whether that was via produce, baked goods, or even native plants. More than a few people came up to us to tell us how wonderful it was that we were growing and selling native plants. The encouragement and support buoyed our spirits and almost made us levitate with joy. We were now one with a farming community and its customers, as well as partners in this important part of local agriculture. We were not exactly sure where we were headed, or which road to take, but we knew we were going somewhere. At Catherine's 40th birthday celebration later that year, a dear friend toasted her and remarked, "You never know where she is going, but it is a fun ride." And it still rings true.

CHAPTER 3:

THANKSGIVING IN TUSCANY— A PRELUDE TO OUR CHEESE DESTINY

The nursery had humble beginnings, but matured into a sought-after regional source for native plants. Over the next several years, we witnessed growth on many fronts: the hoophouse saw an increase in the number of native plants, and more people started coming to the farm to buy plants. It was encouraging to discover the number of people who were interested in how they could convert their yards into habitat-friendly and conservation-minded spaces. With Catherine's two degrees in art and architecture, plus several years' professional architectural design experience, landscape design with native plants was a natural business opportunity. Landscape design and consulting services were a means of differentiating our plant nursery offerings and assisting our clients. We knew that planting native species on our property was just a drop in the ocean. Successfully creating habitat for bees, dozens of native songbird species, plus butterflies, and moths, meant thinking on a larger scale. Helping people establish their own native plant gardens could make a real difference.

The awareness of Yellow Springs Farm grew as Catherine became a frequent speaker at conferences and professional meetings, including the Philadelphia Flower Show. In 2007, *Green Scene*, a magazine published by the Pennsylvania Horticulture Society (PHS) featured the farm story. The article included gorgeous photos and insightful description by PHS President and renowned horticulturist, Jane Pepper. This set a fine precedent, and other publications followed suit in helping spread the word about Yellow Springs Farm. The publicity was priceless and more effective in growing our business than any amount of paid advertisements.

As our conservation landscaping business took root, we found ourselves purchasing attachments for the Kubota tractor that we never knew existed, including augurs, rakes, soil tillers, and a trailer to carry it all. With hired crew members and an expanding list of equipment, we created naturalistic meadows and rain gardens, fortified riparian buffers, and undertook woodland restoration projects, all featuring native plants. Our clients included homeowners and non-profit organizations across Chester County, and as far away as New York City, Maryland, Virginia, and Ohio.

Among our favorite projects was a large-scale meadow at the Mill at Anselma in Chester Springs, right down the road from Yellow Springs Farm. With 250 years of Chester County's industrial heritage, The Mill at Anselma stands as both a historic site and an artifact. The mill retained its original Colonial-era power train, as well as multiple eras of industrial equipment ranging from the late 1700s to the mid-1900s. The mill is still a fully operational grist mill that produces stone-ground bread flour, pastry flour, and dark roasted cornmeal.

As a non-profit organization, the mill's mission included a commitment to reintroducing native plants and meadows on the property. With grant funding available, the board hired us to consult with them on the design and soon thereafter completed two meadows and riparian buffer plantings along the Pickering Creek that powered the mill water wheel. The mill is a treasure in West Pikeland Township. We felt honored and proud to contribute to the history of the environmental improvements within the Pickering Creek watershed. We donated a portion of plants used for the project. A trio of summer interns (Hannah, Colin, and Carrie) rescued Al when he briefly went for an unintended swim during plant installation days; the three made a positive imprint on the entire project. Their leadership, acumen, and commitment to doing the right thing were as inspiring then as it is now.

As neighbors, we were happy to participate in the mill's various events and annual fundraising auction party. When we attended the fundraiser in the summer of 2004, there was bidding on a vacation package that included a seven-day stay at a villa in Tuscany, named La Foce. The photos were beautiful and inspired competitive interest.

With wine in hand, we conferred, and agreed to bid on this. We needed a get-away, and it would be fun to visit our friends in Florence, Italy. Al did not count on Catherine continuing to raise her hand as the bid price increased. Catherine had a very spirited approach and was not going to lose out on the villa. When all was said and done, we purchased a seven-day stay at the villa and made a nice donation to the Mill for its future projects. We invited Michelle and Jim, Catherine's sister- and brother-in-law, to join us in Tuscany. The villa had ample space to fit two couples, with plenty of room to spare.

On Thanksgiving week of 2004, we were off to Tuscany for our stay at La Foce, which means "the meeting place." It is a large estate in the Southern Tuscan region of Val d'Orcia, midway between Florence and Rome, at the confluence of two valleys, the Val d'Orcia and the Val di Chiana in the southeast corner of the Province of Siena. Antonio Origo and his wife Iris bought the estate in 1924 when it was in a dilapidated state. They did not have a background in agriculture, but shared an interest in farming and transforming the land to make it productive and sustainable for farmers' families to live on and prosper. They were keen to embrace La Foce's history of raising sheep and cheesemaking, in particular Pecorino cheese made with sheep's milk.

The Origos restored the late 15th century villa in the 1920s with government financial assistance. The English architect Cecil Pinsent designed the fine gardens. At the time, the property comprised 7900 acres and included 57 farms. The Origos employed 25 families and started a school to educate and ensure the well-being of some 50 local children. They also built 35 dwellings in the 1920s to 1930s for tenant farmers. After the deaths of Iris and Antonio, their daughters Benedetta, and Donata, sold off about two-thirds of the estate and divided the rest between them. Descendants of the family still own the property today and operate it as a resort.

We were fortunate to meet Benedetta when we arrived. We received a first-hand account of the history of the property. Benedetta co-authored a book about the history, gardens, and architecture of La Foce. She kindly gave us a tour of the wonderful gardens. During the history of farming at La Foce, over 5000 acres of wine grapes succumbed to the blight. They sold off that land to another winemaker. Over time, it was replanted with wheat and more grapes. The older olive trees were unaffected by disease and remained productive. In the 1950s and '60s, when many farms were being abandoned in Tuscany, Sardinian sheep farmers immigrated to the area and took over production of the Pecorino cheese. Today, Tuscany is home to many varieties of Pecorino produced by local producers. Aged and fresh Pecorino is widely available both domestically and internationally. This fulfilled the Origo's dream of having sheep and cheesemaking restored on the property.

After spending multiple days exploring different areas of La Foce and its surroundings with Michelle and Jim, we wanted to go off to explore on our own. Michelle and Jim decided to drive to several local towns for lunch and to do some shopping. Without a car, we stayed close and walked to the next town. We started mid-morning, thinking that the next town would be a few kilometers away and we would have a place to eat by 1 p.m. The weather was what you would expect in November, chilly with a mixture of sun and clouds. The winding roads and hills were magnificent and allowed our walk to be a pleasant one intermixed with farms and families going about their daily lives. It was peaceful with the occasional dog barking and farm equipment working the soil.

We originally wanted to eat at a restaurant named Oasis, signposted on a small wooden sign on the roadside. Upon our arrival, we found that the restaurant was "Chiuso Giovedi" or "Closed Thursday," even though our travel itinerary indicated it was "Chiuso Martedi" or "Closed Tuesday." After more hours of walking with no town or restaurant in sight, Catherine remembered that there was a restaurant near the town of Chiarentina. The hills were especially difficult for Al, who was recovering from a torn meniscus. Since it was now past 1 p.m., we kept our eye out for people around their homes, faintly hoping they might invite us in. No invites forthcoming, we hailed a farmer on a tractor. As an Italian speaker since her college years, Catherine learned from the farmer that Chiarentina was another 3 to 4 kilometers away, all uphill, and the restaurant we were looking for was called Il Trinoro.

We were already tired and hungry, and were uncertain whether or not there would be lunch in our future, or if we would find another "chiuso" sign. Rather than take a chance of being disappointed again, we decided to make the three-hour walk back to our villa. Fortunately, there was leftover food in the fridge. Some wine, cheese, and charcuterie were enough to soothe sore feet and tired bodies.

Later that day, when Jim and Michelle arrived back

from their day trip, we compared notes, and then went to Il Trinoro for dinner—by car. As we made our way to the restaurant in the tiny town, we noticed luminaries lining the front of the restaurant and a valet, ready and willing to take our car. After walking for so many hours, we were thrilled to have this service and gladly handed over the keys and entered the restaurant, or so we thought. It did not have the typical guest host station, and restaurant seating was not visible. There appeared to be several dozen well-dressed people on the left side of the room enjoying cocktails and wine. We noticed no tables, so Italy being Italy, we just thought it was an unusual old building and we would figure out the layout, eventually. We investigated the second floor. Most restaurants do not have second-floor seating, but we thought we might find someone who could help us get a table for dinner. Unfortunately, no restaurant staff were to be found. As we made our way back down the stairs to the foyer, our hope was that we would finally be greeted by a maître d'. We were not disappointed. Catherine approached a well-dressed gentleman and inquired about getting a table for four for dinner. He had a serious, and not so friendly look on his face and said, "You are in my home" or "Questa e mia casa." Our dining disaster continued. We were quick to apologize for the misunderstanding and told him that we were hoping to have dinner at Il Trinoro. He said that it was right next door, but he booked the entire restaurant for the Thanksgiving holiday celebration with ex-pat friends from the UK and the USA. The Thanksgiving celebration was in honor of his American friends.

Others, overhearing our story, seemed amused by the

circumstances. Before we left the town, we stopped by to see the actual restaurant. It would have been the perfect place to celebrate Thanksgiving. The staff had set and decorated the tables beautifully for the festive dinner. Not ones to give up easily, we made a reservation for the following day. Third time's the charm, we reasoned.

By the time we left the restaurant, it was 8:40 p.m. We next drove to Montechiello where the only osteria turned out to be closed. Are you sensing a theme here? For a country famous for its food, we were having a difficult time getting any. We headed for Pienza, a larger town in the area. We knew La Luna had been closed earlier in the week, so we figured the odds were good that it would be open by now. As we walked in, by now it was nearly 9:30 p.m., an older woman looked at us from the kitchen in a not-so-friendly way. The optics did not look or feel good, but given the late hour, we knew we were running out of options for dinner. A tall man came out of the kitchen to inform us they were closed and the chef went home feeling ill. The aromas coming from the kitchen were appetizing, so we knew the food here would have been excellent.

Catherine's use of the local language shifted up a gear, complete with hand gestures, almost pleading with them to feed us. Usually, the Italian language and hand gestures get results, but not that night. When she asked for a recommendation given the late hour, he suggested the bar down the street. We passed it on the way to this restaurant, but it had little appeal. Given that most restaurants in small towns close early, we knew this was our very last chance for dinner. Catherine sprinted past us and headed out the door, exclaiming, "To not have lunch is a crime, to not have dinner is a tragedy!" She ran down the street to the bar. By the time we arrived behind her, she was already in discussion with the owner. He said they were hosting the annual dinner for the Council of Pienza Sheep Farmers and the small dining area was not open to the public. Here we go again.

One long table dominated the room. Around it sat 40 sheep farmers wearing work clothes and boots covered with straw, mud, and worn patches. It was a beautiful sight to behold. There was a bounty of food, wine, and a lot of animated conversation and hand motions. Catherine used her best Italian, negotiating skills, and persistence to get us a dinner table set up in the back near the coat closet. We were the only people there besides the farmers, so we agreed to eat whatever they brought—no menu. To our surprise, in honor of Thanksgiving, they served "tacchino"—they had made turkey for the hungry American couples. We gave thanks that after two days of trying to be served a meal in an Italian restaurant, we had not only succeeded, but had unwittingly engineered a date with destiny.

At the time, we gave little thought to the Council of Pienza Sheep Farmers, but Al made a mental note of how he liked the collegiality that is built among a group of farmers raising sheep, milking them, and making cheese. Farming was in their blood and made up their daily lives. Their dress was what you would expect from farmers: rugged, practical and a little worn. Most were in need of a shave and a haircut, but it was for the most part a portrait of perfection. A photo would have been impolite, but the indelible memory it created with the food and wine would never be forgotten.

In 2004, cheesemaking was on one of our brainstorming spreadsheets in an electronic folder, but it was not at the front of our mind. Four years later, when we ventured into cheesemaking, this experience would be an emotional and foundational influence. That night in Pienza, we planted yet another seed that would grow into a community. We would later become the founders of the Chester County Cheese Artisans group. It was not exactly the size of the Pienza Sheep Farmers celebration, but it was a start of building a community of like-minded people working the land to support their community.

When we visited La Foce, we had been at Yellow Springs Farm for three years. Like Antonio and Iris Origo, owners of the La Foce in the 1920s, we were not traditional farmers or experienced at farming. Their experience taught us that, with the will to learn and curiosity, we might also accomplish many of our goals. The Origos had a ten-point plan for the farm. We, too, had a plan for Yellow Springs Farm, even though it was not a ten-point plan. Our plan was amorphous and organic . It would have many twists and turns. The 2004 trip to La Foce and the Thanksgiving dinner with the Council of Pienza Sheep Farmers in attendance was a prelude to our goat cheesemaking destiny. We just did not know it at the time.

CHAPTER 4:

REMEMBER RENA'S BIRTHDAY -
MAY 20, 2006

When we first moved to Chester Springs, our friend Valetta mentioned that she knew a couple who lived nearby and had a bank barn that housed goats and chickens. Chris and Leslie weren't running a farm business, but Valetta thought we would have a lot in common, so she arranged for us to meet for dinner at a local Italian restaurant. All had a good time and when dinner was over, Chris and Leslie welcomed us to visit their farm, meet their goats, and share a meal. Leslie suggested that rather than mow our farm fields, we should introduce goats to provide a more authentic farm atmosphere while reducing emissions from our tractor. We had a vague idea that raising a few goat kids could turn our grazing goats into milk-producing farm animals. We could use the milk for our own consumption and perhaps make cheese on the weekends. At the time, we did not envision an artisanal goat dairy. That would come later when the economic reality of having a seasonal native plant nursery hit home. For now, having fresh milk and making a fresh or aged cheese in our kitchen was an intriguing idea worth further exploration.

We hesitated about the investment in fencing and run-in sheds, but all the commitments and expenses seemed comfortably familiar, given Catherine's experience with horses since she was nine years old. Goats would be just like horses, but smaller, right? We purchased our first goats from Chris and Leslie, cementing an enduring friendship. Rosebud and Dora were both female—two does in goat-speak. This was an intentional choice to avoid growing the goat herd by accident. But it also gave us two chances to have kids eventually born on the farm. After you have one or two cute goat kids, you feel you need a few more. And just like that, we had a "potato chip problem." You have one, two, then a handful more, and then you are in the pantry looking for another bag. There were also the practical aspects to think of in terms of growing the goat herd and having more babies. If we wanted to have milk and make cheese for our own consumption, we would have to continue to breed Rosebud and Dora, since goats go out of milk six to nine months after giving birth. In this case, it meant building a new goat shed and installing more fencing, so could have even more goats.

Rosebud was stout with strong shoulders and a wide girth. If she were to play sports, we would have tried her as baseball catcher, or perhaps a discus thrower. Her black coat with white spots was graphically singular and easy to see in the field, even from the second-floor office window. Dora, a few years younger, was a golden tan. She was photogenic with expressive eyes, tall, leggy, and a bit prissy. She was opinionated about having to leave the shed on rainy days, preferred only the finest stems in each flake of hay, and called aloud in a high pitch when she expected to be served.

Despite marrying later in life, we were content not to start a family, yet we found ourselves drawn to the idea of the wonder and joy that might come with witnessing a birth on the farm. Rosebud was registered with the American Dairy Goat Association, and membership included a directory of registered goat herds in each state. Anticipating her next heat, the fertile hours in a doe's reproductive cycle, the search began for Rosebud's Romeo. After a few phone calls, some confusion of terms and conditions, and time spent with "how to" articles on goat breeding, we hatched a plan to introduce Rosebud to a buck at a farm in the Lehigh Valley, about an hour from where we lived. How do you know when a goat is in heat? The books said she might be ready to breed when she wags her tail or becomes more vocal. After the autumnal equinox in September, the diminishing daylight stimulates fertility in goats. Throughout October and into November, we patiently watched and waited. We were cautioned that she would be ready to conceive for only about 12 to 16 hours once we suspected she might be in heat. If we missed the moment, our next chance might, or might not, come after another 20-day cycle.

Ever the optimists, we called ahead to the farm to arrange our conditional appointment, hoping we had our timing right. When you put a doe in your SUV, and drive on the Pennsylvania Turnpike, you draw some odd looks from fellow travelers who have noticed your unusual passenger.

We arrived with veterinary health documents and a check for fee payment. We introduced Rosebud to Alfredo, the buck. Alfredo is the Italian version of Al, so we chuckled at the synchronicity of the eponymous buck. How would we explain that the sire of the first kids to be born on our farm was Al(fredo?) But it was not a fruitful meeting. Rosebud stood in one corner of the barn while the buck snorted and stomped, but she ignored him. Soon they were both bored, and the experienced herds-people in the group all agreed that Rosebud was not in heat. They warned us that the trauma and upheaval of travel might cause a hormonal change that would disrupt the heat cycle, or that she might be close to entering heat. We learned that we were looking for a "standing heat." If this were to occur, Rosebud would willingly stand near the buck and invite a close encounter.

The trip was not for naught. Rosebud got a dry run for car travel. We went home with a buck rag sealed in a plastic bag. We had rubbed this towel near the buck's scent glands on his head and on his malodorous chest. After 16 days, we were to show her the rag daily and pay attention to the reaction. We could smell the stench on the rag from yards away, so we knew she could too. "Is she or isn't she?" was the question each day. Our spirits dropped when nothing appeared to happen in December, and then December 21, the winter solstice, came and went. As the daylight hours were about to increase, we read that both male and female goat fertility would diminish.

What were you doing on Christmas morning in 2005? We put aside any plans to open gifts, host guests, or cook a special meal. As if getting in the Christmas spirit, we were in our shed admiring Rosebud's convincingly strong standing heat. It was embarrassing to call folks we hardly knew to interrupt their holiday with Rosebud's call to action. But farmers know animals have a different calendar, and every day is a workday, so they kindly agreed to make time for another Rosebud visit that afternoon.

The SUV was ready, and we headed back for a second date. This time we packed a fruit pie, too. If we were to impose on a family's Christmas, a gift was in order. Things were going fine as we pulled into the driveway. Then Rosebud, getting a whiff of things to come, lurched forward from the cargo area, putting a front foot through the center of the pie box. The three wise men we were not.

Our hosts introduced Rosebud to Pepsi, as Alfredo was taking a break from the breeding shed. Things happened quickly. It suddenly seemed so simple, just like the books described. Rosebud was on the way to motherhood.

Goats have a five-month gestation, so we were expecting kids around May 25, 2006. Our breeding calendar was prominently posted on the mudroom clipboard. It was a quick reminder for us on the way out of the house to do barn chores, as we continued counting cycle dates month by month. Rosebud showed no signs of heat in January, so it was likely she was pregnant. Goats show a larger belly, mostly in the last four to six weeks before birthing. We measured her girth every few weeks with a piece of baling twine. Each data point was recorded in a chart, and with marks on the twine. She was getting bigger, but was she just getting fat on spring grass?

As anxious first-time parents, we looked for mentors. Our neighbor Liz had several children and was a nurse in the birthing center of a nearby hospital. These credentials were enough for us, so we asked her to help us determine how things were progressing with Rosebud's pregnancy. Liz came over with a stethoscope, cozied up to Rosebud, and listened for a fetal heartbeat. "Not conclusive," she said, and began palpating Rosebud's jelly belly. With no livestock experience or training, she was working with little knowledge, but Liz's gut instinct was that Rosebud was expecting. Each week made it clearer that kids were on the way, and Rosebud's size implied that it might be a multiple birth.

We called a veterinarian to have Rosebud vaccinated about a month before her due date. He was nonplussed and assured us we would likely find the kids standing and nursing when we came to do barn chores. All the books said a veterinarian was unnecessary for goat births and the only thing we might need hot water for was to make a cup of tea for the anxious human bystanders.

On the evening of May 20, 2006, Rosebud was pacing a bit in the stall. She ate little dinner, and we felt it was the big day. We watched and chatted, hoping our presence would encourage things to move forward. With no such luck, we went back uphill to the house, and agreed to return in an hour or so when we would likely see newborn kids. We found no kids but saw Rosebud's water had broken and she was in labor. She was having contractions and looked rather uncomfortable. With a plant business, we were used to getting our hands dirty on the farm, but this event brought us wide-eyed into the world of raising livestock.

We had a "how to" book in hand for advice, and expected a birth within 20 minutes or so, but nothing happened. More pawing and pacing, standing and lying down, fidgeting and worrying, and finally Rebecca was born. She was beautiful; first of all, she was a "she." We definitely knew it was better to have does for future milk and to avoid the reproduction management that would be needed if suddenly our herd included male goats. Rebecca was black like her mama and had gorgeous white spots that looked as bright and clean as freshly fallen snow. Her velvety black ears were floppy, soft, and so fun to touch. Rosebud licked Rebecca and bonded with her briefly before continuing to pace, paw the ground and irritably deny Rebecca access to nurse from her teat. We knew another kid was looking for an exit strategy, but our first priority was Rebecca's breakfast. With both of us holding Rosebud still, we managed to get some colostrum, the first nutrient rich milk essential to a newborn development, in a baby bottle. As we tried to feed Rebecca an ounce or two of mamma's manna, we found our supermarket baby bottle nipple was far too big for Rebecca's tiny mouth.

As we made our first goat milk delivery to Rebecca's velvety muzzle, we worried about Rosebud and our yet un-born second kid. It was close to midnight when Rosebud finally lurched, moaned, arched her back, and pushed one more time so that a kid somersaulted through the air like a young gymnast just learning her routine. Rena had been poorly positioned in utero as a breach birth—ready to exit butt first. Dystocia caused Rosebud hours of distress. We had no idea how to intervene or assist her, so we watched and worried, hoped and prayed. Then, we rushed to scoop up the runty little goat finally born with a bang. She was a she—so lucky—and she was barely breathing. We wrapped her in towels, as it was chilly in the barn. Rosebud tried to clean her, and we did too. We fed Rosebud some water with grain and molasses, thinking she deserved a special treat and energy reboot after hours of labor.

Rebecca was able to stand and finally nurse, but our second kid could not hold her head up, never mind her spindly body. We called the emergency number for veterinary help and were told they could come first thing in the morning. Around 6 a.m., as the sun was just thinking about rising, a young veterinarian arrived. She was not optimistic. She started IV fluids for the goat kid, injected a few medications, and showed us how to use an eyedropper to get some colostrum into the kid's mouth. The prognosis was that she had less than a 30% chance of living 24 to 48 hours, and even worse odds of thriving. The time she spent turned and folded in the birth canal had been hard on Rena. She was oxygen deprived, and likely had aspiration pneumonia already. Her delay in obtaining colostrum exceeded the ideal one-hour window we were informed about, potentially resulting in a permanent compromise to her immune system. The vet told us the goat would become less able to absorb the nutrients and colostrum components each hour, even if we somehow got her to swallow the golden milk. She had no sucking reflex, she could not stand, her body temperature was unstable, her motor skills seemed delayed, and her ligaments contracted, so her legs could not straighten even if she were to try to stand and bear weight. The veterinarian left us with some tools but not much hope. She told us to call tomorrow if the goat was still alive and she would come back to check on us.

The scrawny black goat kid with white stripes down her face and a heart of gold was precious to us from the moment of her dramatic entry. We called her Rena for reasons both prosaic and poetic. It is common for livestock herds to use the first letter of the dam's (mother) name to continue the lineage in the progeny they birth. Hence Rosebud's kids were named Rebecca and Rena. But Rena's name was also an Americanized contraction of the Italian "Caterina," in honor of Catherine and her Italian roots. Our farm resulted from many moments of serendipity, synchronicity, and chance encounters. This is a great example of a seemingly quotidian moment that might have faded with time, but instead became a pivotal memory that grew in importance as years passed on the farm.

This name was not simply a choice to label the goat after the ego of the farm founder in a moment of hubris. Catherine was born two months prematurely and given very poor chances for survival, let alone thriving. She weighed just over three pounds at birth and spent weeks in an incubator before leaving the hospital. Her mother almost died in the birthing process. At about three pounds, Rena the goat was similar in size. Looking into her eyes, Catherine projected onto her the courage and will to live that Rena needed to survive.

Our neighbor Liz saw the barn lights had been on all night and heard some of the commotion. She suspected the

big birthday had come but could not have possibly anticipated how it unfolded. Eileen, another neighbor and fellow
alumna of Catherine's alma mater, Smith College, heard
the news and brought over a casserole, plus sleeping bags.
She knew Catherine would sleep in the straw for days, pro-
jecting her own zest for life onto the struggling kid. Eileen's
kindness was the beginning of a connection that stretched
over the years as she returned to the farm many times with
her adult daughter and granddaughter, Emmi.

We continued to have Rosebud nuzzle Rena to encourage bonding. Rena still could not stand on her second day,
so we showed Rosebud the eyedropper that contained her
milk and fed Rena while Rosebud watched. Rena looked
weak and glassy eyed. Catherine had the idea of giving her
a few drops of espresso. "If it worked for me; it might help
her," she thought. Please don't construe this as medical advice, but caffeine is a stimulant, and it helped. We developed
a makeshift, ad hoc physical therapy program to help Rena
stretch and straighten her legs. We helped her stand by supporting her with one hand under her ribs. Although wobbly,
she had a glint in her eye when she saw she might somehow
reach a teat and feed herself. Progress was unsteady. Each
minute, things changed. There were hours when she re-
gressed and lost energy. She had trouble pooping, and then
she had diarrhea. There were body fluids of every kind in
that shed, but only one thing mattered—Rena was alive.

You could date the founding of Yellow Springs Farm
Goat Dairy to 2008, when Al and Catherine Renzi got the
first permit to make cheese from the state Department of
Agriculture. But for us, the dairy began when Rosebud gave

birth to Rebecca and Rena. It was the first time we squeezed a warm teat to release fresh, raw milk with its distinctive aroma and color, still at body temperature. Because Rena lived and thrived, we thrived. Everything starts somewhere, and Yellow Springs Farm Goat Dairy roots originated in a goat shed on Christmas 2005.

Even if you missed a few classes in school, you likely know that male goats, called bucks, don't make milk. The females, called does, have teats, and lactation starts with re-production so newborns can feed and grow. Our goat herd was to become largely a herd of does because we were to become cheesemakers. As our herd of does expanded, our knowledge of optimal breeding strategies also grew. Our study of genetics and milk quality helped us understand early on that our success would depend on a few well-chosen bucks, too.

We learned that the University of California at Davis tested goat hair for genetic variations correlated to production of alpha-si casein, an important amino acid present in milk. This called for a quick review of biology texts to refresh our knowledge of Gregor Mendel's 19th century studies of how traits are inherited by combining two gene alleles—one received from each parent. Goats that carry dominant alleles- AA or AB variants- are most likely to produce milk that yields more cheese per gallon of milk. This translates to more productivity for each dollar spent and each hour of human labor, plus much better cheese. With each hair sample submitted by mail to California, we cheered when the results affirmed our goats had AA or AB variants so that their dairy careers were promising from the start.

We also studied the American Dairy Goat Association (ADGA) records that include Dairy Herd Improvement Association (DHIA) data. We saw this correlation with bloodlines in certain Nubian goats. Many people say that milk is a commodity, but we assure you that not all milk is created equal, and not all goats produce milk equally. We needed an excellent breeding sire if we were to continue to breed goats and produce award-winning cheeses.

Goats don't arrive via UPS or U.S. Postal Service. We hitched up our horse trailer and left the farm one evening after milking was done. As daylight waned, we arrived in Frederick, Maryland, where we had reserved a bed-and-breakfast room for a rest en route to northern Virginia. When Catherine called to make the travel plan, the host assured her there was plenty of space to park the horse trailer overnight, but the narrow streets and brick row houses lining the historic area suggested otherwise. Al had gained invaluable experience in backing up the trailer in tight spots, often in the dark, while dating Catherine. In fact, one might say his success and perseverance were a prerequisite. Al attempted to back the trailer in between two brick walls with only inches to spare, causing traffic to be stopped for what seemed like an eternity. The neighbor stood on his porch watching to see if his property was about to be damaged and loudly announced that he was a lawyer. Nice guy! The trailer parked without harm, but everyone's patience was thin. We left early the next morning, before the lawyer was out and about. The rest of this story was much warmer and happier.

After driving a couple of hours, we arrived at a lovely farm in bucolic Loudon County. We were so fortunate to connect with Mary and Bob Clark and Cherry Hill Farm, where we met Marco. He was calm, tractable, and a bit smaller than some, so he was easy to handle. More importantly, he produced lovely does. They were logically also on the smaller side. Some made lots of milk, and others were bound for homesteads because of low production. Goats with less capacity to be dairy stars were sometimes sold as pets, or companion animals for anxious Thoroughbred racehorses. As the Yellow Springs Farm goat herd grew, we culled and selected the herd carefully. It was not much different from fielding a team for baseball. We often joked that we related so well to the movie *Moneyball* where the Oakland A's management used statistics and analysis to put together a winning team despite a lean budget.

Over the years, with the same attention to genetic traits and confirmation, we selected breeding bucks from the offspring born on our farm. It started with Dante, then Mick, Retro, Maestro, and later Rembrandt, and Charlie. All the bucks we cared for, and appreciated, but Dante was a failure in the squad. He was eager to reproduce and had genetic attributes, but he also was too eager to break through the gates, head-butt humans, and threaten those who crossed his path. We bred for temperament in addition to milk quality, so Dante ended up in our freezer. Al would barely eat the meat, as his bitter memory of the difficult decision to take one of our goats to the butcher seasoned the flavor of the meat. This was the first time, but not the last; we brought goats for meat processing. Chili and curry softened the edges of goat meat's strong taste, real or perceived, and this chapter concluded with more lessons learned about running a commercial dairy.

A turning point in our farm came with the purchase of Gareth from a goat breeder in Oregon. Her Nubian herd had years of excellent milk records celebrated in several countries. The genetics were outstanding, and there was a waiting list for this breeding stock. Gareth arrived at PHL— Philadelphia International Airport—on United Airlines. His itinerary included a change of planes in Salt Lake City. We followed his flights and connections throughout the afternoon and then went to the freight terminal around 10 p.m. to meet our newest buck. The monitor said the plane landed, but the staff told us the goat was not on the manifest. Our hearts sunk and voices quaked. We knew this young animal had to be tired and thirsty, and we were sure he was there somewhere.

There were inaudible two-way radio conversations with lots of static, as queries were made around the airport. Finally, we heard "Wow, he has long ears" and we knew someone was looking at our Nubian goat in his crate. When the cart and trolley came into view, the freight handler was beaming. He had never seen a goat like this before, and everyone thought that Gareth was adorable. We opened the crate door to pass through soda bottles with red nipples and Gareth quickly latched on. One was filled with water and electrolytes, and another with milk. We checked Gareth for bumps and bruises, and took his temperature. All seemed in order, as he had traveled amazingly well, and we headed back to the farm. Farmers don't stay up past midnight for parties, but they do stay up late when kids are born, or in this unique case, when goats fly in from the West Coast.

There were many bricks in the solid foundation we

formed to support Yellow Springs Farm. The cornerstones were the carefully selected bucks and the thoughtful match-making—which doe and which buck would be likely to produce the most productive offspring. But equally important were the connections we made with a web of friends, especially Valetta, Chris and Leslie, Liz, and Eileen. Thinking back, we realize how these ties were instrumental in getting us off the ground. We would come to build many treasured relationships along the way as we continued on our farm journey.

CHAPTER 5:

CONNECTING LANDSCAPES
AND FOODSCAPES

W e thought about making goat cheese even before we started the native plant nursery in 2002. Perhaps it was more than a coincidence that at least three different bosses in corporate life gave Al or Catherine Spencer Johnson's 2002 best-selling book, *Who Moved My Cheese?* Did they subconsciously sense that change was on our minds, or were they hinting that they thought it would be better for us to change our career paths and lifestyles? Maybe we just accidentally went to work wearing barn shoes a few too many times and co-workers thought we smelled like bad cheese.

The book's parable describes the characters looking for cheese in a maze to find nourishment and keys to happiness. In this context, cheese is a metaphor for what most everyone wants in life. The book reminds all to stay flexible, nimble, and good-spirited in the face of change. It was not meant to be read literally, but in some ways, it is the seminal text for what aspiring cheesemakers must learn.

Why not start a goat cheese business during the 2008 Great Recession? A year before Bear Stearns and Lehman Brothers crashed over the mortgage derivatives business losses, Wells Fargo bankers turned away Yellow Springs Farm business credit, citing the high risk of lending to a farm business. Risk assessment is in the eyes of the beholder, but history shows perhaps their judgment was clouded.

Having a back-up revenue plan was important in case the recession lasted longer than expected and the negative impact on the farm and plant nursery increased. With a financial crisis, we expected the large ticket landscape installation projects we enjoyed until then might become less common. After all, people have to eat! We thought that the goat cheese could be a good complimentary revenue stream to diversify the farm enterprise and allow use of our native fruits and plants to make unique farmstead goat cheeses. What do two serial entrepreneurs do in the face of crisis? Of course, we start another business.

The cheese option came more into focus at that point. Even with the risks of capital investment and fickle market interest in specialty cheese, we felt it was the right thing to do. We also looked ahead at the orthopedic wear and tear of digging countless holes, moving heavy materials, and bending and lifting for hours each day. Landscape projects would likely need to be phased down to smaller scale work, even if there had been no recession. We had an idea to revisit the concept of making goat cheese. It sounded like fun. Dairy use paid homage to the farm property's dairy history. It was a cow dairy for most of the 19th century and still included the historic bank barn, springhouse, and farmhouse where the Himes family worked. We walked in their shadows, and felt called to respect, honor, and continue the fading dairy heritage of Chester County, PA, and specifically the farmstead we had the privilege to own.

Our initial inquiries and research about starting a dairy at the farm were not promising or productive. There were naysayers at each corner, and they told us what we could not do, why it was a bad idea, and warned us of all that could go wrong. Things changed one morning when we were milking and Brian from PASA (Pennsylvania Association for Sustainable Agriculture) appeared in the driveway accompanied by personnel of Pennsylvania Department of Agriculture, Bureau of Food Safety, Division of Milk Sanitation. In short order, the momentum changed direction, and the discussion turned to "why not?" They offered to take an unofficial milk sample back to the lab, just for informational purposes. A few days later, the report arrived to prove that our raw milk was of high quality and exceeded metrics for sanitation. No wonder it tasted so good.

As our first doe, Rosebud, began lactation in 2006, we hand-milked her with a stool and metal bucket in a run-in shed. There was straw on the ground, other goats milling about, and we only sometimes used gloves. Minutes later, we enjoyed the raw milk in our coffee and cereal, and we stored the rest of the raw milk in repurposed glass juice bottles in the kitchen refrigerator. By the weekend, we needed to use the milk before wasting it, and hence we made our first fresh cheese. It was a spreadable chevre, but had so much more flavor and soul than commercial cheeses from the supermarket.

In the early years, our home cheesemaking cultures and kitchen cheesemaking tools largely came from New England Cheesemaking (www.cheesemaking.com). Al gifted Catherine a countertop cheese press for Christmas. We tried unsuccessfully to repurpose a metal bread box to a cheese-aging container. Al's academic background in microbiology came in handy as we accumulated thermometers, pH meters, pipettes, and more to embrace the science of cheesemaking.

We drank raw milk and ate raw milk cheese. Our friends enjoyed the kitchen counter cheese batches, even though the products came with warnings of our steep learning curve. Wine and beer pairings shared likely helped set the tone of jovial satisfaction, even if the cheese quality varied.

When we first started cheesemaking in our kitchen with leftover milk from the week, it was a simple soft cheese recipe that was easy to make. We would heat the milk to room temperature, add a cheese culture, and then let it sit overnight. During that time, the cheese cultures consume the sugars in the milk and produce lactic acid, which makes the milk more acidic, curdling it to produce cheese. The next day we would drain the curds in a cheesecloth and let that sit for several days. Then, we added salt, herbs, or other flavorings we had on hand before consuming it on toast.

We didn't use a lot of science, food safety measures, and technical processes during these cheesemaking sessions.

We were just making cheese in the kitchen like you would prepare bacon and eggs—simple, straightforward, and fast. The hearty reception for cheeses we produced for friends and family encouraged us that this was something we could do. In the realm of the art and science of cheesemaking, our cheeses first fell into the "art" category. They reflected our love of food, local food, and simple farm-to-table preparations. Looking back, those first small batches were among our most exciting cheese feats. Imagine the sight, sound, aroma, and feel of creating delicious soft cheese from our own goats' milk just hours after we gathered it from the warm teats. It does not get better than that.

With the encouraging first milk test results, we began to map the road to commercial cheesemaking. What licenses and permits were needed? Where could we find used stainless steel dairy equipment? How would we bring our 19th century dairy barn into the 21st century and still maintain its character? The goats seemed on board, and so these were some things that dreams are made of.

We spent 2007-08 working to identify the best path forward. Each challenge helped us improve and refine our practical business and marketing strategy, and gain technical knowledge about cheesemaking. Given the space limitations of the stone barn, we pondered how best to layout the cheese aging room and milking parlor. We needed a good financial plan in order to justify the investment in farmstead cheesemaking. The evaluation process started with the venture proposition to make edible products, such as fresh and aged cheese from raw milk, from Yellow Springs Farm goat milk, with a minimum of capital investment. The next step was undertaking a detailed SWOT (Strength, Weaknesses, Opportunities, and Threats) analysis of Yellow Springs Farm and its capacity to diversify into a dairy-related enterprise. Once that analysis was complete, Yellow Springs Farm had the chance to create a new mission statement with goals combining their plant nursery and milk production businesses into a unified and profitable entity that would also benefit the community. This was perhaps the first time we imagined how to connect landscape and foodscape in tangible ways. No matter how Yellow Springs Farm was going to evolve, we knew we wanted to follow the path of small-scale processing, producing high quality fresh and aged cheeses with a good cheese-aging cave, and maintaining a reputation for high-quality standards.

Our cheese business feasibility studies included farm visits in Pennsylvania, Massachusetts, Vermont, and beyond. We coordinated our itinerary with Larry, our project consultant for the USDA Value-Added Producers grant. Larry held a Ph.D. in food science. After decades of professional experience with the Hershey Company and other food manufacturers, he then worked with the Penn State Cooperative Extension. His assistance with the grant, helping us do the research, and creating a schedule with the cheesemaker visits was invaluable. The Value-Added grant offered us matching funds to research our future business.

Years later, we learned how unusual it was for a small farm to apply for and receive the grant monies. We spoke to the U.S. House Committee on Agriculture at the Capitol about the grant process and offered suggestions for increasing small farms' participation in the program. It remains the only time we ever saw the inside of the Capitol. It is unclear if any change came from our efforts, but the visit was personally memorable. There were staff members, but no elected representatives, in our meeting. This was our chance for a firsthand view of how the federal government works, for better or for worse. We felt like characters in the 1939 film *Mr. (and Mrs.) Smith Go to Washington* and then hours later we were back milking goats on the farm.

We traveled with Larry throughout Pennsylvania during the winter of 2007-08 and met a handful of cheesemakers milking both cows and goats. They shared their experiences getting into the cheesemaking business and plans for surviving. Some were even thriving, but others struggled. We typically met Larry somewhere in central Pennsylvania in the early morning hours. The days were long, traveling miles extensive, but all well worth the trouble. During our visits, the ambient temperatures were 20-30 degrees F. The cheese rooms, aging caves, and dairy barns were damp, and not that much warmer. As we made rounds, Catherine casually said, "I could eat a pork chop right now." Al and Larry smiled at the thought. Cold weather brings out the desire for comfort food, especially with Catherine. Reminiscent of her time spent in Italy, in winter Catherine craves rich meats, and Ribollita (pronounced ree-bohl-LEE-tah), a classic, hearty Tuscan white bean soup (stew) with vegetables, thickened with day-old bread. There was no Ribollita to be found, but we were in farm country, so pork chops were a possibility. As the two-day trip wore on, Catherine repeated her desire for a pork chop. This time, Larry and Al piped in that she was making them hungry, too. There were few restaurants in the area, but it was clear we would have to do something about this.

Al had already learned the lesson years earlier, when Catherine had a craving for beer and curry. There was an Indian restaurant in Malvern, Pennsylvania, about seven miles from the farm that we frequented. Al was not up for beer and curry that night, but later found out that his recalcitrance had consequences. When we were going to sleep, Al leaned over to kiss Catherine, but Catherine would only said, "Beer and curry." Our friend Casey heard about this marital tiff and offered advice to Al. "Never deny a woman her cravings." Back to central PA, we finally found a pub open in a small town. What was on the menu—Pork Chops! Catherine satisfied her craving and kissed Al good night. Lesson learned.

Stainless steel equipment is the biggest investment made when starting a dairy, as we consistently saw with all the cheese visits. Milking parlor, cheese room, and aging room equipment needed to be purchased for Yellow Springs Farm. New equipment was not affordable, so we perused the *Lancaster Farming* newspaper and other publications looking for refrigerated milk tanks, cheese vats, draining tables, cheese racks, and triple bay sinks. We visited dairy equipment dealers in Lancaster County, Pennsylvania, such as EZee Milking, and Fisher & Thompson. Many cow dairies were going out of business because of the low commodity price for fluid milk, the aging dairy farm families, and the temptation to sell dairy farmland to developers. This surplus equipment was everywhere, and attractively priced, but unfortunately, its size and scale would not serve our nascent goat dairy. Ironically, smaller, specialized dairy equipment is much more expensive than used cow dairy items. This was a classic case of supply and demand in our market-driven agriculture economy. We learned a few years earlier that this same economic reality applies to shopping for a small-scale tractor that fits in tight spaces—more expensive, and lower supply than the larger ones. There was a consistent theme in our small farm plans—equipment costs were high and often over budget. The economic reality of making a profit at a small scale was challenging. Spreadsheets and college degrees were helpful, but there were few case studies or role models to follow on this singular entrepreneurial journey in artisanal dairy.

We were always on the look-out for bargains during our PA farm visits. We met an older couple who was planning to retire from cheesemaking. Their son Josh did not plan to continue with the cheese business. He aspired to become a Pennsylvania Department of Agriculture inspector who would call on cheesemakers quarterly to ensure they kept their premises clean and their cheesemaking equipment working properly and within safety guidelines. Goats make only two to three quarts of milk per day most of the year. The better does will produce a gallon of milk during May and June when daylight hours are long, and weather is moderate. It is important not to hold the milk for more than three days before using it for cheesemaking. In early years, we were getting about four gallons of milk per day with a small herd of goats, so we had only about twelve gallons on the third day to make cheese. We needed a very small pasteurizer. Josh's family had a twelve-gallon pasteurizer that would be a perfect starter vat for us, given the number of goats we were milking. This was our first dairy equipment purchase.

Pasteurization is not just a matter of heating milk on the stovetop or in a soup kettle. A pasteurizer in a facility operating under regulatory permit has to include a chart recorder that provides a written record of the temperature and timing, plus a specialized thermometer that measure both air space and milk temperature. Stainless steel is mandatory and the welds must be approved and inspected. This equipment is an investment. Pasteurization prevents unwanted pathogens, such as E. coli bacteria, that could cause food safety issues. Our first pasteurizer was immediately put to good use.

When deciding which cheese varieties to make, we leaned towards those ready to sell shortly after production. This assured good cash flow and also avoided the need for extra inventory storage spaces. The category includes soft cheeses such as chevre, bloomy rinds similar to Brie or Camembert, mozzarella, and fresh feta. Cheeses aged less than 60 days prior to sale need to be made with pasteurized milk.

We had moments of comic relief during the cheese research trips. The unexpected happened, and things did not proceed according to plan. During a trip to the Vermont Cheese Trail, we also sought to fill a landscape client request for a heritage Sugar Maple tree. The Yellow Springs Farm nursery did not have a big enough tree to fit the bill. Since Vermont is known for maple syrup, we hoped to pick out a sugar maple tree while visiting cheesemakers. We visited nurseries and finally came across one the right size and prayed it would travel in the Ford Expedition. We picked up the tree and found it barely fit even with the top branches over the dashboard in the front seat, running right down between the driver and passenger areas so we could hardly see each other, or the mirrors, through the foliage, but the fragrance was pleasant. One complication was that we were only halfway through the cheese trail. How could we get the tree back to Pennsylvania alive and with all its parts intact? We called ahead to confirm our B&B overnight reservation and mentioned, "Oh, by the way, do you mind if we bring an eight-foot tree?" The owners had never had that kind of a request before. They seemed amused but accepting of our situation and offered to put out a hose out so we could water the tree. While on our way back to Pennsylvania the next day, we noticed a car being pulled over by a police car, apparently giving someone a ticket. Unaware, we zoomed by without moving over one lane, which is the law in Vermont when an emergency vehicle is on the side of the road. They had not enacted that law in Pennsylvania that we knew of, but that was not an excuse for being on the wrong side of this vehicle law. In just seconds, a police car with flashing lights came into sight, requiring us to pull over. The officer asked rhetorically why he pulled us over. We said we did not know. Since we were not speeding, we were unsure why we were stopped. The officer was not amused by the tree, especially since Catherine peered at him through the leaves. The officer explained the traffic law, but did not issue a ticket. He gave us only a warning so we could get on the way with our good luck charm, a Vermont Sugar Maple.

We gained technical knowledge on our farm visits, but what we remember most were the emotional connections to the concept of family members working together to create a business, loving the land, and wanting to be a part of their local food community. Yes, there were certainly discussions about which breed of goat was best for cheesemaking, what equipment and suppliers were best, which culture houses that could provide the needed cheesemaking ingredients, and the various pathways to regulatory and marketing commercialization. We embraced the grit and creativity that it would take to do something that we knew was the right thing to do in our hearts, but not necessarily the best objective financial business decision to make for the good of the farm. We had been through a lot of ups and downs running the native plant nursery and we knew if we were to move forward, it would take all of our business experience and creativity to make a go of this venture.

To get our facility and equipment approved for commercial sales, we had to pasteurize three batches of milk successfully with recordkeeping, make goat chevre with that milk, and demonstrate that the cheese was free of pathogenic bacteria. We were so eager to enter the marketplace. The requisite three batches of milk and cheese testing took over three months. Once we completed all the testing, the day arrived in September 2009 for the final facility inspection. The inspector looked at the test results and temperature charts, checked the calibration of the thermometers, and reviewed our application paperwork. Then he inspected the cleanliness of the milk parlor, cheese room, and aging room. It seemed to go on forever, but we received the official seal of approval that allowed us to start selling our cheeses.

We began selling our farmstead goat cheese in 2009 with the dairy manufacturing permit in hand from the state of Pennsylvania, and the Chester County Department of Health approval to sell and sample cheese at farmers market. We had the West Pikeland Township building permit to renovate the dairy barn. Our farm had an ID number from the federal government. The PA Department of Agriculture certified our on-farm lab for milk testing. Our goats and farm premises were inspected, too. Nationwide Insurance representatives had visited the farm as part of their audit in providing our property and liability insurance. Yellow Springs Farm had a sales tax license, and registrations and inspections for five farm vehicles and trailers. There were binders of recordkeeping and paper trails of all kinds. We often quipped only half-jokingly that nuclear power plants were likely regulated less stringently than our little goat dairy. We had worked in financial services and pharmaceutical industries, so the idea of functioning in regulated environments was not unfamiliar to us. Every bit of life experience was helpful to prepare us for cheesemaking. It was unexpectedly the most complex undertaking we ever knew in personal or professional life before 2009.

Once we obtained our cheese production permit, things got a little more complicated. We had to have cheese recipes, flow charts, food safety plans, allergen plans, product recall plans, FDA approved labels, thermometers, pH meters—I think you get the point. And this was just the beginning of all the record keeping, testing, and reporting that we would have to do. It was exciting because even though we were not making cheese in our kitchen anymore; we were producing small batches of cheese with a twelve-gallon pasteurizer, making our production manageable. We could still make a high-quality cheese without the burden of producing large volumes. Years of planning, building restoration, herd growth, equipment purchasing, and testing cheesemaking techniques passed quickly. This arduous process reached a pivot point when we found ourselves back at the West Chester Growers Market selling goat cheeses. This was a full circle moment for us, since it was also the first place we sold the farm's native plants in earlier years.

Making small cheese batches frequently allowed our food creative juices to flow. Initially, we focused on fresh spreadable goat cheeses in various flavor combinations. Everyone had a favorite, but three varieties, plain, pepper and garlic, and Italian herbs, were more popular. Our first hard cheese, Fieldstone, was born in our 40-gallon cheese vat that was not a pasteurizer, but instead enabled us to make raw milk cheese. With raw milk cheeses, the cheeses had to be aged for a minimum of 60 days before selling.

We wanted to create a hard aged cheese that would reflect the "terroir" of our fieldstone barn building dating back to the 1850s. The concept of terroir is commonly brought up in the context of winemaking. In cheesemaking, terroir applies as well. The pasture soils, grasses, water, milking parlor, cheese room, and aging room all have affect how a cheese is going to look and taste. The fieldstone walls of the aging room contained yeasts and molds that were unique and couldn't be reproduced in a more modern facility. We thought why not create a cheese that was the essence of

Yellow Springs Farm and of the historical place in Chester County that we inhabit? As cheesemakers, we think of ourselves as curator. The goats work hard to produce milk from the rich pasture soils and grasses. In the milk parlor, we wear gloves and prepare the teats and goat udders to ensure that we do not contaminate the purity of the milk produced by the goat ladies. Once this milk is collected in refrigerated tanks, our job as the cheesemaker is to not screw up all the work and effort that has gone into the production and collection of this special milk that contains mostly water, proteins, fat, minerals, and the non-quantifiable ethos and terroir of the farm. With minimal cultures to acidify and curdle the milk, we would make the cheese, drain it, mold it, brine it, and then wait for 60 days as it matures. It is a beautiful, dynamic process that is constantly changing after the first week or two. Once the salt brine penetrates the surface to reach into the core of the cheese, the surface molds from the cave begin to grow on the rinds. At first it looked a little scary, but we kept turning the wheels, brushing the cheeses, making sure the humidity and temperature stayed in the 90% range and 50 degrees, respectively. Then, we waited the necessary sixty days required for selling a raw milk cheese. At day 61, we tasted the cheese and for the first time realized what Yellow Springs Farm terroir was all about. The chemistry and microbiology of the farm made the cheese into something very special, and Fieldstone was the first flagship aged cheese for our farm.

On our 2008 trip to Italy, several years after our 2004 Thanksgiving in Tuscany, our last stop was a small restaurant in Florence's Oltrano area, found just a few blocks beyond the Ponte Vecchio bridge after a stroll through the Piazza del Duomo and Piazza della Signoria. It was a place Catherine remembered as stylish and casually elegant from her time studying abroad in the 1980s. As a student, she never indulged in a meal there, but now it was a perfect choice for our special dinner to cap a lovely stay in Italy. We asked the server if they had a local after-dinner digestif to cap off the night. He suggested a black walnut liqueur called "Nocino" which in Italian means "little black walnut." In our effort to develop creative Yellow Springs Farm cheeses that were "of this place," we realized we had many walnut trees on the property that would be a good source of fruit for eventually making Nocino in Pennsylvania.

Most of our experiences with black walnut trees are the green golf ball size walnuts that drop in the fall. They have a black dye that easily stains everything, which is considered a nuisance. We remember collecting dozens of them prior to our Open Farm days in the fall so that our customers would not trip or step on them. The Nocino liqueur does not use the nut itself but requires the use of the fruit before it turns into a nut. In Pennsylvania, it was necessary to produce the Nocino around the 4th of July when the fruits are just about developed. After harvesting the fruit, we would cut the walnuts in half and add them to a glass jar with grain alcohol. We would place the jars in a south-facing window of our house for 20-30 days, after which time we filtered and bottled the Nocino. We would experiment with various spices, adding sugar, and cutting it with water, so every batch was slightly different. During our first bottling, we tasted the Nocino to make sure the flavors were what they should have been, but

Catherine tasted more than she realized. Needless to say, there was more than one "oops" when she poured but kept missing the bottle. We took her off the tasting crew, but it was fun to go through the process together. During the summer, we aged the Nocino and later used it to marinate our homemade hard cheese. We named the cheese "Nutcracker," and it became one of our bestselling seasonal holiday cheeses. Walnut fruit picking became our annual ritual in late June. We would enlist neighbors with black walnut trees to let us know if their tree was bearing fruit. They would either pick them and deliver them to the farm, or we would bring our ladder and harvest the fruit ourselves. Our community neighbors were our first cheese collaborators.

It was a very normal Sunday evening in July 2010 when we were in the office paying bills and discussing the calendar obligations for the week ahead when the phone lit up with incoming messages. Congratulations were arriving in rapid fire to share that we had just won first and second place awards at the American Cheese Society Competition in Seattle, Washington.

Although we had shipped cheese via FedEx about ten days earlier, it seemed like ages ago, and we almost forgot about the competition amidst the many farm responsibilities. This cheese competition was an abstract idea for us. We really could not imagine what it looked like to have about 1500 cheeses appear in one place, and then judged for technical and aesthetic qualities. It seemed like a variation of figure skating competition freestyle scoring, but perhaps more arcane. We almost did not enter because the entry fees were substantial, and it was so expensive to ship the cheeses priority overnight to the West Coast. Our Nutcracker cheese won a blue ribbon, and Red Leaf, an aged cheese wrapped with sycamore leaves marinated in red wine, came in second among goat cheeses. Both were original farmstead creations made with native plants from the farm. It was a dream come true because our message of connecting landscape to food-scape, and a plant nursery to a goat dairy, finally seemed to be validated, heard, and applauded rather than doubted. The importance of these awards became clearer with each new customer contact and media coverage opportunity that followed. Being awarded for our cheese's authenticity, qual-ity, and flavor was a full-circle moment for our farm ven-ture. It was a pivotal point to further encourage us to build on what we started.

CHAPTER 6:

MAKING CHEESE OUR WHEY

L et's take a few steps back to the months before we won our first American Cheese Society awards in 2009. We can trace much inspiration for this accomplishment back to our trip to Italy in 2008.

Farm life has a way of being all-consuming. It is draining to work long hours without a day off for months. When daylight wanes after the September equinox, goats naturally make less milk or even go completely dry for the winter until the next spring birthing season. The plant nursery was seasonally dormant after the first frost, too. This cyclical slowing of nature's paces offered us an antidote to avoid burnout. Traveling to Italy was an adrenaline-filled farm getaway, so energizing that it might refill the reservoirs we drained during the growing and milking seasons. We tried to make "a big trip" every few years a tradition once we learned that, when work felt especially taxing, our wanderlust was a very effective motivation tool.

A global financial crisis was in full swing. Clients were either deferring landscape projects or cutting back the scale and scope dramatically. We had had a good year selling plants and doing landscape design projects to that point, but the change in the economic climate made us nervous about our future ability to make a living from producing items that fell under the discretionary spending column. With worries galore, and strained minds and muscles, we were surprised when it was suddenly time for our next journey abroad.

Our destination was the Italian Alps, unsurprising because of our mutual love of Italy. Catherine lived in Florence, Italy, for two years during her college days, and has thorough command of the Italian language and culture. Since we were both dedicated to making excellent cheese, this wouldn't be a typical Alpine idyll. Our goal was to visit a mix of professional and peasant farmers making tradition-al goat cheese. This sounded a lot better than watching the stock market go down 900 points per day! The United States elected Barack Obama president just three days before we left for Italy. We watched the election results until finally the news was in that Obama would be our next president. This was a momentous time, no matter what your political preferences. We packed our bags and left on a hopeful note, overhearing the campaign slogan "Yes we can" on broad-casts, headlines, and chatter everywhere we went. This phrase became a fitting theme of sorts for the trip.

Despite Catherine's linguistic skills, she had little luck

creating an itinerary for visits to Italian goat cheese makers while sitting in our home office. Our luck changed when she found a website for "Mama Margaret's Cooking Adventures." This did not exactly seem like a good fit initially, since their business was focused on lovers of food and wine. After writing to find out whether they would be able to assist us, it became obvious that Margaret, based in Canada, had never put together a goat cheese tour before, but she seemed up for the challenge. Margaret made inquiries for a few weeks and then proposed a rather impressive itinerary of visits to goat cheese makers in the Piedmont region in the northwest of Italy. Many know this region for Fiat factories, together with industrial and financial centers in Turin. In contrast, foodies like us focus on Piedmont for its black truffles and beloved wine varietals, including Barolo and Barbaresco. She clearly had our number because the well-rounded plans also included a local wine tasting, truffle hunting, and several cultural visits. After ironing out a few details, a brief phone conversation convinced us that Margaret truly existed beyond a website, so we sent our deposits.

It is hard to explain what needs to be done before you leave five goats, four cats, two dogs, a few thousand plants, and four repair-prone old buildings behind for a working vacation. We felt comfortable only after writing eight pages of reminders and helpful tips for the three capable women who generously agreed to care for the farm during our trip.

All preparations complete, we both slept soundly on the first leg of the trip to Paris, and then we flew on to Turin. Just as Mama Margaret had promised, our guide Elio and his partner, Silvia, met us at the Turin airport, complete with a cardboard sign with our names in black marker. We made quick introductions, and together set out for the one-hour drive to our accommodations in Alba.

As we drove, the conversation was going well, with all the requisite animation and hand waving. Catherine quickly transitioned to using her Italian language skills fluently, but Al prepared a few basic Italian phrases so he could participate in the conversation. Blissful days immersed in Italian culture and language often involved easily learning common phrases such as "*Allora, Cosi, va bene,* and *Fa niente.*" These are comparable to short English phrases such as "Oh well," "What can you do?" or "That's the way it goes." These words fit many contexts. When paired with emotion and body language, they were enough for conversation. We soon discovered that our very modest pensione hotel room for the week had sparse furnishings but was conveniently located one floor above a bakery and coffee bar. Whatever it lacked in charm, it more than made up for with wafting morning aromas of custard, espresso, and flaky, sugar-coated crusts.

We had been traveling almost 24 hours, so we planned to have a quick rest at the hotel, then a brief dinner within walking distance. But when Silvia invited us to her home for dinner, we found it hard to say no, despite our fatigue. Our motto on the trip was: "If not now, when?" as we agreed to meet in the hotel lobby at 8 p.m. Almost everything about our farm life started in more or less this same way, with a "carpe diem" moment.

We felt very welcome in Silvia's home, even though we were among strangers. The distance between strangers and friends closed quickly. Silvia and Elio kindly invited a few other guests too. Among them was Franco, a friend of Silvia and Elio, who travels extensively for business. He spoke English well. Between Franco and Catherine acting as translators, Al gained an understanding of the conversation. Hospitality needs no language. Convivial laughter and multiple courses with perfect wine pairings meant the three-hour meal flew by. Our household is no stranger to long dinners, but we were out of practice during the growing season from March through October. Regrettably, goat births, planting schedules, and taxing farm chores often compromised our dinners at home.

The meal began with glistening olives and assorted dried sausages, thinly sliced with other cured meats. We are not sure if cheese was served, but if it were, it failed to impress us because our attention was held by an appetizer specialty of the region, *vitello tonnato*. Silvia layered slices of bread with cold sliced veal and then covered them with a creamy sauce flavored with tuna. She proudly made this using her grandfather's recipe. She explained with care each ingredient, including herbs, spices, and anchovies. Her recitation of every step of the preparation was precise. Silvia was an architect, so her focus on fine details and patience to curate small matters others might overlook was apparent in how she prepared the food, in the chosen table settings, and textiles, too. There was a pasta course featuring silken, slow-cooked ragu, a roasted meat course, and homemade pudding (*budino*) for dessert. Wine flowed generously. We were tutored in the differences between *Barbera d'Alba* and *Barbera di Asti*, and then enlightened about the various sub regions for each. We compared wines from the sunny side of

the valley to those made with grapes grown on windy hill-
sides. This was a working trip, and we loved our work.

Fun fact: the "Slow Food" movement was founded
in 1986 by Carlo Pietrini in Bra, less than 10 miles from
where we enjoyed our lengthy repast. Most Italians we met
gently scoffed at the celebrity of Pietrini and Slow Food
International. They found its principles and ideals inherent to life and unoriginal. They know no other way to enjoy
food—the concept of fast food never having made inroads.
McDonalds, KFC, and other chain exports are popular in
Rome, Milan, and other large cities, but are a world away
from Alba. Whether we sat down to dinner at 7:30, 8, or 8:30
p.m., the dinner would end in exactly three hours. After a
week of noticing this, Al began to observe carefully to see if
anyone looked at their watch to determine that three hours
was up and it was time to leave, but he never found that to be
the case. Perhaps Italians have it built into their DNA, or they
have a specific circadian rhythm at this latitude. Regardless,
the hospitality, food, and conversation were fabulous, and it
was a great way to start our cheese journey abroad.

The next morning, a Sunday, came quickly, and we were
on our own. We headed to the city's seasonal outdoor market in the square. Cheeses, wines, and other local food items
from farms and artisan producers filled the tables and tents.
Although we were fond of the West Chester market in PA,
this atmosphere was grander. The scale was much bigger,
and the mood was joyfully chaotic. We learned, for example,
that hazelnuts are everywhere in Piedmont in November.
This was our first exposure to an Italian farmers market,
and it was a kid-in-a-candy-store experience. We wanted to

buy everything, especially the cheese. After perusing each of the stands, we purchased a book on Italian cheese. There were cheeses of every shape and size piled high on tables. Some were round, and others were in block form. Some were oozy and ripe, while others shined with crystals, bragging that they were suitable for grating. In the U.S., the cheeses were required to be wrapped, refrigerated in coolers, with sample pieces in container on ice for food safety purposes. This market had each cheese unwrapped, with wonderful cheese aromas mixing with the smells and colors of the charcuterie and vegetables offered by the other market vendors. If you wanted to try a sample, the cheesemaker would take a wheel and slice a good chunk for you to taste. The cheese rinds were rustic, multi-colored with no special rind treatments or packaging. This type of aging allowed for the best of the cheese flavors to come through in both the rind and paste of the cheese. We were on a goat cheese mission and were determined to see, taste, and bring home as much as we could. Be careful what you wish for. By the end of the trip, we avoided eye contact with anyone giving away cheese samples. But, on this first day, our disposition was more open-minded and our ambitions to learn about cheese were infinite.

We took about six or seven cheese varieties back to the hotel, but we had no refrigeration in the room. At first, the room smelled great—it was so appetizing. By the second day, the aromas were strong, but still mesmerizing. By the third day, the cheese smelled like old, wet shoes. Little did we know that the housekeeping staff was monitoring this, as well. When we arrived back to our hotel after our cheese excursions on the fourth day, we realized that we did not smell the scent of any cheese in the room. We found this curious, so we started to look for cheese everywhere in the room but could not find it. We discovered that the house-keeping staff simply hung our plastic sack of cheese outside, using the shutter hardware, to keep it fresh, and keep the room from becoming overwhelmed with odors that ranged from ammonia to sulfur to decaying stench.

That same Sunday happened to be the last day of the annual truffle festival in Alba. The night before at dinner, a guest who arrived late, but in time for dessert, was kind enough to give us tickets to the festival events. We had no idea what to expect but proceeded to the convention hall. The mixed aromas of the salami, truffle, and wines wafted through the hallways. The experience was not only sensual, it was heavenly. We were almost overwhelmed by all of this food, but we were fit for the challenge and its opportunities.

We had limited knowledge of black truffles. They are expensive, hard to find in the USA, and an uncommon delicacy. Some high-end restaurants in Philadelphia promote their truffle specialties and we have occasionally tried them, but they have never been a significant part of our palate or our lifestyle. In Piedmont, truffles are an integral part of the culture. They are perhaps even more prized than artisan cheeses and local wines. Not knowing the history and prevalence of truffles, we were unprepared for what we saw there. Truffle hunters lined the aisles with their temperature-controlled mini-display cases that resembled humidors for fine cigars. Each vendor's table included one or more glass cloches covering carefully selected truffles. Bigger and uglier

truffles were apparently better. Scales tared to the milli-
gram were in view; ready to negotiate the very expensive
purchase transactions. A quirky bunch, the hunters trans-
formed into merchants as they awaited prospective buyers.
The typical attire was a strange mix of styles suggestive of
Peter Pan, Rip Van Winkle, and Yogi Bear and included a
laced waist jacket, broad belt, high-crowned hat with feath-
er, red stockings, and boots. We surveyed their wares from a
polite distance. We found it all a bit intimidating. It felt like
being at a Sotheby's fine art sale or racehorse auction where
we kept our hands still and eyes averted for fear of accidentally purchasing something we did not want and could not
afford. Black truffles are a serious business.

The previous evening at dinner, our hosts asked if we
had an interest in trekking. The only thing we knew about
trekking was the Star Trek-related trekking. We could not
imagine that we were going to meet Captain Kirk and have
Scotty beam us up to the Starship Enterprise, so we asked
them what trekking was all about. It turns out trekking is
the same as hiking, so we felt assured this would not be
complicated. Having the foresight to bring boots and walking shoes, we jumped at the chance to take advantage of the
good weather being outside. After spending the morning
touring the farmers market and truffle festival, we agreed
we would meet in the afternoon for a trekking adventure.

Elio and Franco met us promptly at 2:30. We stopped
at a bar for a coffee or espresso, which is standard operating procedure in Italy, before undertaking anything, even a
phone call or buying a bus ticket. As we gathered with others, we soon learned Elio was not just participating, but he

was a leader of the trekking club. His low-key, modest nature was again apparent in not telling us in advance about his role with the club. The immediate benefit of being the guest of a "Super Trekker" was a 50% discount for our tickets. There were about 40 other trekkers in attendance. We saw men and women of all ages, and children, too. Some groups seemed to be three or even four generations of a family together. We had no idea how all of this was going to work, but we boarded a bus to our hiking destination. We learned during the introductions that this was going to be a nine-kilometer adventure. It was maybe more than we hoped for, but why not? We were soon walking through a prominent chestnut forest. November is the time of the chestnut harvest. The shells of spent nuts littered the forest floor, most likely as a result of a feast for the local squirrels and other mammals.

In the U.S., it was, and still is, very rare to find a large stand of chestnut trees due to the chestnut blight of 1904. This fungal disease of the bark devastated the species and by 1940, the American chestnut was virtually extinct. Given we were born in 1957 and 1964; we had never seen a mature chestnut woodland. This was a thrill for two nature lovers, and a real inspiration for our woodland restoration work at Yellow Springs Farm. There are ongoing efforts underway in the U.S. to commercialize blight- resistant chestnut strains, which are often crosses between the Chinese and American chestnut varieties. It will be many years, if not decades, before we see the results of these efforts in America.

At the end of this long trek, an open pavilion filled with picnic tables set with modest tablecloths, local wine, and cheeses greeted us. Or were we hallucinating? Could there really be homemade bread slathered in local honey, and an open fire with chestnuts in a roaster? This being Italy, yes it was real. The pavilion was in an elevated area, so we could look out over the hills as the sun set and see the towns below begin to glow. As we wrapped up this impromptu party, everything was lit by twinkling stars.

We were tired at the end of the trek, but there could not have been a better way to start the trip. What a gift for our first full days in Piedmont—the market, the truffle festival, the trek, and the sunset feast! None of the first two days' activities were planned or written in Mama Margaret's official itinerary. But we had already amassed more accolades and praise for our hosts' hospitality than we have space to share.

The next day was our first "workday" for cheese visits with our guide, Ettore. We met Ettore the night before at a second marvelous dinner hosted by Silvia and Elio. We were immediately comfortable with his affable, easygoing style. It was a bonus that he spoke English fluently, too. Ettore is a very fit gentleman in his 50s. He would be our driver and companion over the course of our week in Piedmont. Our first day of the planned itinerary started promptly at 9 a.m. In introductory chatter, we learned Ettore was not originally supposed to be our guide. He was a last-minute replacement for a woman named Carol. We did not know Carol either, but this change caused us some pause. Ettore openly said he had never done this job before, and knew little about cheese, but he seemed to be willing to learn with us. To set the tone, Ettore explained he was born and raised in Pavia, a small city between Turin and Milan. He was a retired math teacher and school administrator who had lived for many years in Egypt. Together, we embarked on what would be a fascinating journey into the world of cheeses while building a friendship around discussions of politics, film, and a hundred other topics.

Our first stop was at a husband-and-wife goat cheese-making operation. For those not well versed in the Italian road system, the rural areas tend to have very narrow and poorly signposted roads. Ettore informed us he did not know the exact location of the farm, but, once we were close, we would stop at a church suggested by the farmers. He would call them, and they would meet us with a car and escort us to our final destination.

The farm owner, Sylvia, arrived in her car and we followed. We felt the excitement of starting our adventure and learning about goat cheese. We met a roadblock, however, in the form of a local farmer coming down a steep hill on a tractor with a cart of hay in tow. It was clear that he could not get past us, or we him. So now what? The roads are typically not that wide. It is hard enough to have one car traverse the narrow roadway let alone two vehicles, one of which was a very large farm tractor. After a lot of hand-waving and incompressible Italian dialect, the farmer agreed to back-up and allow us to pass. It was no small feat on this steep hill. On either side, there was not much space to maneuver and given the hilly nature of the pathway, it was extremely dangerous to venture either too much to the left or right. Finally, after what seemed like hours, the farmer managed to safely remove his tractor out of the way so that we could get by. At this point, the fact that the winding road that we continued on was not paved and full of potholes seemed like a gift compared to what we just witnessed with the tractor. Without having Sylvia to follow, we would never have found their farm. Maybe that is the way they like it here. All we could think was welcome to Alba!

Our kind hosts showed us their farm, animals, and cheesemaking facility. Since this was our first visit, we were very excited to see their process of cheesemaking. November is not the best time to see cheesemakers at work. Goats typically dry out slowly or stop giving milk completely at this time of year. The goats that we learned about on this trip, Camosciata Alpine, are particular to the Piedmont region and parts of France. They are a fairly large breed with a brownish coat color. The typical good milker can give up to one gallon of milk per day. This is similar to the goats on our farm. Although our Nubian goats, a breed originally from equatorial Africa, had longer ears and a different coat type, there were many similarities between the two types.

This visit was particularly interesting in that the couple had never met an American before. They were educated professionals with careers off the farm, but they preferred to focus their world view on Europe and Russia. Although initially apprehensive about meeting us, they felt more comfortable given our shared values and interests. They baked bread to share during our time together and served lunch before our cheese tasting progressed. While we were eager to learn the art of cheesemaking, we realized making connections with people and learning about their lives is as important as a cheese recipe.

Our next stop was another hilltop village, where we met

a goat cheese maker named Mario. On the road to his farm, we saw protest banners opposing the municipality's plan to bring a highway closer to this village, where time seemed to stand still. We were unable to see the actual milking and cheesemaking process because of the lateness of the year. However, Mario had a few cheeses aging from earlier in the fall.

The grape harvest had recently ended, and we noticed a basket filled with grape must, the waste after grapes were pressed for wine, on the cheese kitchen floor. When we asked him why that was there, he mentioned that he had cheese rounds in the basket soaking in the must. This was our first, and by no means our last, lesson in using what you have to make a great cheese. This moment of inspiration was perhaps the time and place where our Red Leaf cheese began.

Mario's farm was rustic and included historic stone outbuildings and a typical Italian farmhouse. Fruit trees were growing all around the property, with persimmons ripe for picking. The milking area was located outside in a three-sided shed area with a roof. There was no real sanitation here, unlike what is required by U.S. regulations. Yet, the family and staff were all still alive and thriving and producing great cheese.

The dirt floor, lack of running water, and open air gave us pause. It was a contrast to all the expensive and time-consuming regulatory and food safety hoops we needed to jump through to obtain a cheese permit in Pennsylvania. For example, we needed to have a concrete floor, a drain for wash down with hot water, washable walls and surfaces, a sanitary procedure for cleaning the goats' udders, wash sinks to wash your hands and a three-bay sink to wash, rinse and sanitize all the mandatory stainless steel equipment for milk storage, pasteurization, and cheese production.

Our early farm system evolved from hand milking like Mario (prior to our selling cheese commercially) to having an automated vacuum milk pipeline system that would enable one to milk the goat and then have it automatically transferred to a milk tank that served as a milk refrigerator. Our tanks were small (50-85 gallon) compared to what a cow dairy would have (250-500 gallon or more). Mario, on the other hand, did not have refrigeration for milk. He just made cheese after each milking. It does not get any fresher than that, and it is probably why he did not have any contamination issues. The cheesemaking process is all about fermentation and acidification, which is anti-bacterial by force. Cheese was actually first made by shepherds as a means to preserve milk. Cheesemaking did not start as gourmand's indulgence or a high culinary art.

Mario's cheese room was a lesson in minimalism. He had a few cheese trays, a draining table, a pot that sat on a propane burner, a few cheese molds, and that was it! Nothing fancy, nothing extra, everything had a use. The cheeses were simple, aged only briefly, and most were freshly eaten in days without weeks or months of *affinage*, or *aging*, in an underground cheese cave.

You might be wondering why we would want to visit these types of rustic cheese operations, using methods that would not pass U.S. regulatory food safety standards. There are several reasons. First, we wanted to see how small-scale farmstead operations functioned. We were also interested in learning how to make particular traditional Italian style goat cheeses such as Robiola.

Robiola is a small, round goat milk or mixed milk cheese which is made predominantly in the Piedmont region of Italy. It can be made with 100% goat milk or mixtures of two or three milks: goat, cow, or sheep milk. We love rich, creamy, bloomy rind cheeses that most Americans associate with Brie or Camembert. Robiola fits the bill. The fresh version of Robiola can be eaten within days, but most of the Robiola that you see in cheese stores is aged at least 10 to 14 days before it is released. The inspiration for what would later become Bliss and Kennett in our Yellow Springs Farm cheese line ultimately came from Robiola. But we are getting ahead of ourselves, again.

Mario's small scale cheesemaking and goat milking inspired us. We realized that similarly we had to work within the fixed footprint of our ground floor barn area. It had to serve as a milk parlor for goat milking, a cheese room for cheese production, an aging room for aging cheese, and a retail space for selling cheese. All of this needed to fit into a space of about 400 sq. ft. "Impossible," you say? Well, yes, we actually said that, too. We needed to figure out how to make the disadvantage of a small historic building work to our advantage. We were looking at Piedmont's examples as best practices, and ways to model this rustic cheese we were hoping to make.

Mario and all the cheesemakers we visited knew how to produce cheese in a small space without spending a fortune. It was good to see how they planned their cheese-making space, what equipment they used, how they milked their goats, and how they packaged and sold their cheese. It was a real-life demonstration to prove that you do not have to spend a million dollars to start making cheese. Some cheesemakers spend a million dollars on a cheese space because they can, or perhaps because they aspire to a larger scale operation. In contrast, we realized we wanted to have a community-based farm that focused on the cheese, not the fancy space and equipment. We wanted a boutique farm operation that would allow us to have a high-quality, unique product. It seemed quality over quantity was the way to go, although that concept would always challenge us throughout the years when we needed more revenue to pay the bills.

After a good tour of Mario's farm and the one-room cheesemaking facility, we were invited to stay for lunch. We had told Mario that we would like to learn how to make Robiola cheese—just like the cheese that he was making. He said he would be happy to help us, but he had to prepare lunch first. When you are invited to lunch in Italy, it is something to look forward to. We met several American WWOOFers from Worldwide Opportunities on Organic Farms (WWOOF). This is a worldwide organization to link visitors (WWOOFers) with organic farmers, promote a cultural and educational exchange, and build a global community conscious of ecological farming and sustainability practices. WWOOF started 50 years ago and has grown from a small group in the 1970s to a worldwide community of hundreds of thousands of people today. We also met several itinerant Italian farm helpers who rode in on horseback. Mario referred to them as the "cowboys." This colorful group of people was all there to help Mario run the farm.

One little glitch in our visit, or at least for Al, is that Mario only spoke Italian. Fortunately, Catherine could speak and translate Italian, allowing Al to take notes. Our lunch lasted several hours. We did not even discuss cheese during that time. Mario wanted to talk about politics, art, and language. On a side note, he mentioned to the cowboys that Catherine spoke better Italian than they did. They looked a little perplexed by the comment but seemed to take it in stride. Good food and wine will do that.

Finally, after three hours of eating and talking, we asked Mario if he could specifically help us with making Robiola cheese. He agreed that now was the time. He carefully gave us the process for making the cheese. It was fairly simple in retrospect, but at the time, we felt like we struck gold. After we completed taking copious notes, he stopped and looked at us and said, "I am happy to give you the recipe for making Robiola cheese. But you will never be able to replicate our goats, our milk, our climate conditions, or our aging process like we do it here on our farm and in the Piedmont region. There is a certain terroir that makes Robiola what it is. "I suggest you go home and make your own cheese," he flatly counseled.

That gave us pause because we realized that as much as we like Robiola, it might be better to appreciate what has already been produced in Italy as Italian cheese, like Italian art, and Italian language. We would not think twice about reproducing or duplicating Italian art or the Italian language, so why would we want to duplicate Italian cheeses. What if we could go home and be inspired by our Chester County farm in Chester Springs, PA, and make cheeses that is of that place? It is probably the best lesson about cheese-making that we ever learned, even up to today. It is a lesson that you would never obtain from a book.

As we were winding down our trip to Alba, our last stop was dinner at a typical local trattoria restaurant. This particular place did not have a menu. They just start bringing food out to your table. As you finish one course, another arrives. This went on for at least four courses, and it got to where we both were feeling like we had enough to eat. Catherine asked the server if there was going to be more food coming. The server responded, *"C'e sempre qualcosa di piu."* This Italian phrase means "There is always something more," and we would come to find that life on Yellow Springs Farm epitomized this thought. Yes, there is, and will be, always something more.

An Air France pilots' strike extended the downtime during the return flights from Europe. The return trip began from Turin to Florence by train. After visiting our dear friends from Catherine's undergrad years, also known as our Florence family, we had a booking on an Air France flight to Paris. Upon arrival at the tiny Peretola airport, we found long lines and anxious disorder everywhere. Catherine overheard rumors of a strike and canceled flights. She decided to move into the shorter line for "Club" passengers. We did not know which Club, but why not? We needed answers, and folks ahead of us in our former queue were getting nowhere. When our turn came, the Club agent did not ask for membership or credentials. She was nonchalant in saying that she could get us on a flight to Paris that day, but not to the U.S. She said the airline would pay for our hotel in Paris for a day or two, or until the strike ended and connections began to flow better. Catherine tried to stay calm. She locked eye contact and exclaimed that we had "babies" at home waiting for us. Those were goat kids, but that's a detail. Italian women are generally very smitten by young children, even if they have no inkling of actual motherhood. The mention of babies started a flurry of phone calls and two-way radio messages. The agent left the counter and returned with a female colleague wearing a reflective orange vest and carrying a flashlight lantern. Perhaps she was doing airplane maintenance on the tarmac minutes earlier? The two conferred, typed fervently on two keyboards, and then grinned sheepishly. "We will get you home to the babies," they said. They handed us boarding passes to Brussels, with a connection to Boston, where we indulged in New England clam chowder while waiting for our final flight leg to Philadelphia. We arrived without our luggage, but that was hardly a surprise given the changes and connections. The zigzag of change laced with a little luck and kindness of strangers crystallized for us that the travel journey, like our lifestyle at Yellow Springs Farm, was not an ending, but just one more chapter in our story.

Our plane arrived late at night, with us being jet-lagged and exhausted. Our loyal farm sitters reassured us that all critters were well, and no problems occurred in our absence. We woke up on time the next morning, looking forward to seeing our babies, and starting morning chores only to find the dogs uncharacteristically refused to eat breakfast, and our best milking doe had wood splinters in her udder. At this point, we realized that we were back to real life and were again on call. Life at Yellow Springs Farm was never boring, and our roles, in contrast to times in corporate cube life, never felt unproductive or underappreciated.

CHAPTER 7:

COMMUNITY SUPPORTED AGRICULTURE

S oon after we moved to Chester Springs, PA in 2001, we
met Sam, who was running a Community Supported
Agriculture (CSA) program a few miles away. It was inspiring to see this acronym in action. People came together to
learn about the origins of their food, support a local farm,
and occasionally volunteer to assist with farm projects, harvest days, or staffing the share room where members came
weekly to pick up their food. We saw firsthand how factors
such as garden pests, soil health, and weather events all contributed to, and sometimes detracted from, the quality and
quantity of food in our farm share. We willingly shared the
upside and downside of these risks with the farmer. After a
year or two, we joined a second CSA at Charlestown Farm.

We were deeply affected by the experience and knowledge shared. We were inspired by the passion and purpose required to raise local food. Our great-grandparents would have found it surprising; years ago, it was normal to grow your own food and purchase things you didn't grow from your neighbors. Everything old was new again; nationally, CSA programs thrived and grew through the 2010s.

When we received our cheesemaking permit from the PA Department of Agriculture in fall of 2009 and immediately started working on building a Goat Cheese CSA for the 2010 season. There was no model or specific prototype for this, but the inspirations from numerous vegetable and fruit CSA programs were easy to grasp. Before long, our goat cheese CSA had share pickup sites in four counties. Several vegetable CSA farms offered to collaborate and serve as pickup points for our cheese share so that members could pick up cheese and veggies at one stop. Our close connections with produce growers inspired seasonal cheese varieties at Yellow Springs Farm. If there is a tomato season, then why not pair it with Goat-za-rella seasonal goat cheese? If saffron or black walnuts are available for just a few weeks each year, that is cause for an expression of this bounty in a new cheese recipe. The small batch, local market, and presold nature of CSA shares made this sort of creativity and invention possible. Without CSAs, our cheese creativity would have had to endure long lead times of distribution chains and new SKU approvals processes bogged down in layers of management at larger retailers.

CSA trends began to change in 2016-18 across all regions and product categories. Customers preferred product on demand instead of waiting for pickup dates. They were growing accustomed to home delivery championed by Amazon Prime. Special diets such as nut-free, gluten-free and vegan became more common, so people expected customization of their food choices instead of simply sharing in the farm's offerings of the week. Our cheese CSA morphed, too. We began to partner with larger, regional food hubs such as Common Market and Lancaster Farm Fresh that had CSA pickup sites in several mid-Atlantic States. We retracted our direct cheese CSA shares pickup sites to a more hyperlocal focus in Chester County. By 2020, almost all our cheese shares were being picked up at Yellow Springs Farm as we gradually transitioned away from other pickup sites.

There were unexpected benefits to this full circle return to hyperlocal cheese shares. We spent more time getting to know our CSA members and their families. "Friends of the Farm" was an unofficial moniker for the people who stuck with us through sleepless kidding seasons, power outages, and cheese batches that were simply not as good as we hoped. Our CSA members were a special part of this equation. They paid for their shares months before the first cheese pickup, so we would have cash flow in winter months when goats produced far less milk, and our native plant nursery was dormant. Some CSA members became farm volunteers; some even became part-time employees. However, there was something even bigger that is harder to describe. The CSA members became an audience, a fan club, and a reason to start again when all else failed. The CSA members cheered loudly when we won cheese awards, shared photos of their celebrations with our cheese, and offered ideas for recipes and pairings that made our cheeses shine in home kitchens.

Our cheese played a featured role at members' family weddings, graduations, birthdays, holiday celebrations, and special occasions of every kind. It was witness to many marriages, divorces, births, and deaths in our farm community. We shared moments of angst with parents of adolescents, and empathy with adults caring for aging parents. Cheese share pickup days were occasions we looked forward to because we loved hearing about college admission applications status, a better job opportunity, a new puppy, and accomplishments in team sports. In turn, our CSA members wanted to know what we named the goat kids born last night, see the latest news coverage featuring our farm, and hear about endless farm equipment repairs. The cheese CSA continued for 12 years.

We had no extended family nearby in Pennsylvania. Yellow Springs Farm was the opposite of a multigenerational farm with children, grandparents, cousins, and assorted in-laws chipping in, coming and going, and filling gaps when things needed doing on the farm. Ours was a husband-and-wife farm business. The CSA members became a community that sometimes felt like family.

We found one plus one equaled three most days as we complemented one another's skill sets—Al in science and analysis, and Catherine in creativity, animal care, and motivation to make the Farm run. That said, many times even "Three" were not enough. The farm benefitted from the invaluable contributions of over 100 volunteers, interns, employees, and a few WWOOFers. Eric was a WWOOFer who bicycled to the Farm about three hours round trip each day for about a month. Then he worked without a break in the goat dairy. His fortitude and friendly attitude were inspiring. It was a pleasure to offer him lunch each day, despite the impressive number of calories required to fuel his vigor. No leftovers went to waste. We often invited volunteers, interns, customers, and employees to a group lunch table in our home or patio. These were opportunities to learn and build friendships. Sharing food remains a joyful and effective way to break the ice and connect with others in memorable moments. What we gave up in personal life privacy, we received in exponential returns of professional conviviality and singular emotional connections. The pizza parties using our own cheese, farmstead herbs and tomatoes, all cooked in the patio wood-fired cob oven built using mud and straw from the farm, were times to cherish.

Opening the private sphere of our home to the farm and its people became the norm. The precedent was set in 2002 when we were asked to feature the property for the Chester County Day tour. This charity event benefits Chester County Hospital. When over 1000 guests walked through in a few hours, their enthusiasm numbed and thrilled us. We capped the day with a reception for about 20 neighbors, and never again considered privacy in the same vein. The farm was meant to be shared. On a subsequent Chester County Day, we gave each visitor, about 1200 people that year, a packet of native wildflower seeds. It was a pleasure to inspire others to love nature, and a few guests sent us photos of the blooms the following year.

As our native plant nursery inventory grew, we started

having Open Farm Day two or three times per year. Our first Open Farm Day felt like an audition for a play or a tryout for a sports team. We set the date, prepared with adrenaline, and waited. Would anyone come? Would anyone care? In 2002, we had no goats to bring people with the cuteness factor, and no food to bring people to sample farm products. We only had a nascent native plant nursery, a 19th century farmstead with newly established gardens, and tons of enthusiasm for the future. People came. They talked, they shared, they asked good questions, and they bought some plants. We collected email addresses and postal addresses. For the first eight or ten years, we mailed hundreds of yellow postcards to announce Open Farm Day. We stamped them and applied address labels in the den after dinner for a week or two until the pile was ready to mail. We welcomed non-profit organizations such as Longwood Gardens, Pennsylvania Horticultural Society, Natural Lands, Audubon Pennsylvania, and Trout Unlimited to partner with us.

Open Farm Days were much more than revenue opportunities. They were progress reports on the farm's growing footprint and changing business practices. Open Farm Days were the perfect time to launch new items softly, such as gift certificates or organic plant care products. Open Farm Days were networking events where we met people from so many businesses with shared missions, and we met future customers who would eventually hire us to design native landscape features for their properties. Sometimes we met media professionals who later wrote feature articles or created videos featuring our farm work. We remember at least two times our loyal visitors came with freshly baked cake to keep our sugar levels and spirits high throughout the weekend.

It was an enormous effort preparing for Open Farm Days, especially in May of each year. Every plant needed a name tag, a price tag, and an information sign. Repeat customers knew there would be gaps in this process, but they came back anyway. What we lacked in marketing, we tried to make up in personal service, human connection, and attention to the plants—each nurtured by hand, weeded by hand and selected especially to make our customers' gardens more eco-friendly and aesthetically pleasing.

Interns from West Chester University, Penn State, and Temple University were often keys to the spring chaos and event preparation. We welcomed their interest in sustainability and plants. Sean was a few years older, and about to be a landscape architect when he helped at the farm. He was keenly interested in learning, and among the most productive students who we worked with over the years. He shared that besides his educational goals, he came to the farm with a pedometer. In pursuit of his future wedding date, he set a goal to walk at least 10,000 steps each day towards his fitness and weight loss targets. Within the first days, he was surprised that he exceeded his daily step count before lunch. In hindsight, we would have made much greater profits if we had marketed the Yellow Springs Farm Weight Loss Plan, but instead we continued to focus on plants and goat cheeses.

Our two dear friends, Georgie and Judy, were long-time checkout staff at our cashier canopy. They proudly wore the bright yellow Yellow Springs Farm T-shirts that indicated to visitors that they were the ones who knew how things worked, and where to find Catherine and Al for more specific answers. They made things hum as they understood how to *guesstimate* the price when all else failed, make sure the customer was pleased, and keep eyes open for wandering children, parking jams, and plant wagons rolling downhill without a human escort. Al disliked wearing his Yellow Springs Farm T-shirt, so for him we decided to start offering bright green baseball caps with the farm logo. People in the know were not fooled, and would find Al working on fern propagation, or irrigation system maintenance. They knew he was intentionally quiet, but a wealth of information.

"Horses in the driveway" was a phrase we coined on Open Farm Days. It captured disbelief, perseverance, resourcefulness, and a hint of disgust, all in a few words. There were plenty of Open Farm Days when the field flooded, the dog ran away, or staff helpers called out at the last minute—quotidian moments. Only once were we busily preparing to welcome our annual community reunion when Al raced up to the nursery to tell Catherine there were horses in the driveway. The goats were amused but not distracted from eating their hay. The horses were agitated and lost, and were precariously close to the road as commuters' rush hour began. Lawn maintenance crews, with their loud equipment scurrying across the way on Meredith Lane, hardly noticed these 1000-pound lawn ornaments close to the road.

Catherine spent decades caring for horses before she and Al purchased Yellow Springs Farm. Al assumed she would hurry to help and have the horses safely under control quickly. But the daze of Open Farm Days preparation, and the exhausted hangover of goat kidding season dulled Catherine's sensibility that afternoon. She walked down the driveway, got a glance at the horses, and turned around to head back to the nursery in denial of what she saw. There was no time for this unplanned event. It was so absurd, we had to laugh, but in the moment, we were too overwhelmed to enjoy the humor. As Catherine started walking back to the nursery to resume the Open Farm Day's preparations, Al looked on in disbelief and said, "We have two horses in our driveway. We just can't pretend they are not there!"

Once we regained our senses, we tried to fashion horse halters from baling twine. We looked for goat feed to use as treats to lure the horses uphill, away from the road. When we asked the lawn guys to turn off the machinery that was spooking the horses, they mostly pretended not to understand us. We tried to slow down traffic, knowing the horses were moments from bolting. One car stopped, and the driver announced she was a veterinarian. "Ok great, so can you help us? Do you have a halter or lead rope?" Catherine asked. "No," she said and drove away.

A few cars somehow thought that honking their horns as they passed the chaotic scene was a good idea. This was not helpful in calming horses. When we saw an SUV pulling a horse trailer coming, we thought our rescue had arrived like a star on the dark horizon. The driver stopped, and we asked if she could help us get the horses in her trailer for safety while we figured out where they lived. She thought about it for a minute and then told us she didn't want her trailer to get dirty and was worried about the risk of disease or contagious conditions these unknown horses might leave behind. She, too, drove away.

While all this was happening, our next-door neighbor had a great idea. He was a dog owner with no horse experience, but he solved the problem by getting in his car and driving slowly down the road to follow the horse poop piles. He came back to tell us he found the broken fence rails about a mile away. With one car leading the way, driving very slowly, and another vehicle behind us warning drivers of the special procession, we walked the horses home. By then, the horses had become more trusting, and probably tired. An arm wrapped around each horse's neck with four fingers crossed over the head above the nostrils served as the halters we did not have. We had only carrots and baling twine to maintain control. As their pasture and barn came within sight, the horses exhaled, and so did we. We settled the horses in stalls in the empty barn and filled their water buckets and gave them each a flake of hay. Then we wrote a note on the barn blackboard and left our phone number. No one called us, but a week or so later, someone had repaired the fence, and the horses were again in their pasture. Whew!

Still today, when life is so absurd that truth seems stranger than fiction, we shudder to imagine what else might happen and when the next shoe will drop, we turn to one another and know it is a "horses-in-the-driveway" moment.

Once goat cheese entered the farm offerings, we added "Holiday Celebrations with Cheese" to the calendar, in addition to spring and fall Open Farm Days. We welcomed guests for a few days each December for cheese tasting, cheese sampler gift boxes, various colorful goodie bags with bows, and complimentary hot cider. Some young goats enjoyed, or at least tolerated, wearing fleece antlers and Santa hats. For a few years, we made goat milk caramel sauce, goat milk soaps, and even goat cheese chocolate truffles. Together, these were unique, seasonal gift ideas. Shopping at the farm was more fun than the mall or Internet experience, except that it mostly happened outdoors. We owned lots of wool socks, insulated pants, gloves, and parkas. One December holiday celebration stands out, but we don't remember which year it was. We were very wet and cold because it rained and snowed. Shivering and sneezing, we finally moved all the activities into our house. The holiday music was playing, and the fireplace roared with logs ablaze. Only a few guests came because the roads were icy and our gravel driveway made all-wheel drive imperative. But those who came stayed for a while. It was an impromptu party. The business side of Yellow Springs Farm melted into the personal presence again, and we were all better for it.

2021 was the final year for Open Farm Days. The last farm event in September was anti-climactic because so much had already happened over twenty years. It would have been greedy and unrealistic to expect it to be better or bigger. Open Farm Days, even in the ending year shadowed by Covid cautions, were almost perfect reunions. We loved seeing generations of families come together to the farm, especially on Mother's Day weekends. We watched grandmothers teach grandchildren how to talk to plants, and children buy a particular plant for their parents to recall a special childhood memory. People selected trees for memorial gardens and wedding gifts. We welcomed women who had been visibly pregnant in past years, and now holding a child. The circle of life and nature's intentional, unwavering rhythm gave sanity and meaning to the frenzy of preparation and hard work over the seasons. Human life cycles were so clearly intertwined with the cycle of growing plants from seed, and then using cuttings or root divisions from the source plant to start even more plants. Some plants died because of human error, or mice, or an unexpected pest. The nursery moved forward with more plants, anyway. The farm community mirrored this resilience as it grew and changed, but it continued no matter what. It was community supported agriculture at its best.

CHAPTER 8:

IT TAKES A VILLAGE

Many of the same customers who lifted our spirits and nudged our native plant nursery forward were there to embrace our cheeses. Brand managers had warned us that this business model—plant plus cheese-- was flawed, and that we had to choose to focus on one thing or the other. They were wrong. Business jargon might describe this pattern of cross-marketing to a customer segment. In practice, everything happened more organically, no pun intended. As the farm product offerings expanded, we had more opportunities to get to know our customers better.

It was fun to sell plants, but selling food-edible products added a dimension of love to the equation. In the Italian traditions of our parents and grandparents—food is love, and love starts with food. We considered adding "love" to the ingredient list on our product labels, but the regulatory requirements for text and limited space for anything beyond the basics quickly sobered us.

Our passion for food is personal. Cheese is a great icebreaker, and it helped us build friendships with chefs, journalists, community leaders, and store owners. People of all ages loved our cheeses. Kids were brutally honest and asked for seconds. English was a second language for some customers, but the universal vocabulary of pleasure was all we needed to know. French, Spanish, Portuguese, Italian, and visitors from countries afar were smiling as they enjoyed our goat cheese.

The small batch nature of our production allowed for a lot of creativity in producing novel cheeses. There was no reason to copy from other cheesemakers or to simply follow recipes that you can find on the Internet. It became our mantra to look around the farm and find something that could make a great cheese. We always say that it is easy to make a good cheese, but hard to make a world-class cheese.

During our first three or four years of production, we developed multiple seasonal cheese ideas, including Pepito with peppercorns, Mellow Yellow made with saffron, Purple Passion with lavender, and the list goes on. We developed almost 30 varieties of fresh, bloomy rind, and aged cheeses.

Our two most popular cheeses were bloomy rinds: Cloud Nine, a white quarter pound fluff ball; and Black Diamond, an ash-coated cheese inspired by the traditional *Valencay* cheese of France. We thought many of the bloomy rinds on the market were bland and tasted more institutional rather than reflecting a farmstead production. These cheeses were very labor intensive to make because we shaped and molded them by hand. With the increase in demand for these two ripe cheeses, our production inched towards resembling more of a manufacturing operation rather than a small creamery business. But given the importance that these two cheeses made to our farm profitability, we kept adapting ways to make the process of creating at least 300 to 350 Cloud Nine and 150 to 200 Black Diamond per week more efficient and less labor intensive.

Producing a larger quantity of cheese for key customers called for established cheese recipes, procedures, and processes. Sometimes it was necessary that the art and creativity of cheesemaking take a back seat to the science, food safety, and packaging. It was a challenge to assure that customer orders, shipping, and off-site events did not take away from the core values that got us to this point in the first place—our love of food and community.

We were community focused, and so inspired by Judy Wicks' 2013 writing about her career path in the book *Good Morning, Beautiful Business: The Unexpected Journey of an Activist Entrepreneur and Local-Economy Pioneer* which focused on her work at the White Dog Café in Philadelphia. We had been acquainted with Judy for many years before Yellow Springs Farm. Catherine remembers Judy was occasionally a classmate in the Wharton School's Small Business Development Center evening classes for entrepreneurs long before we even dreamed of owning a farm. Then, in an earlier entrepreneurial chapter for all, Judy purchased Catherine's architectural necktie designs for her Black Cat boutique on Sansom Street in West Philadelphia. When Al was offering cheese samples at a Philadelphia natural foods cooperative in Mt. Airy, Judy was there signing books. Soon after, we attended Judy's talk at Ursinus College. We left with a signed copy of her book, but also a wealth of inspiration. Judy might not have realized it, but principles gleaned from her business made our business better.

Publicity matters and everybody loves a winner. With the desire to spread the word about Yellow Springs Farm, we entered the American Cheese Society competition each July beginning in 2010 until the Covid pandemic changed the plans for large conferences and gatherings in 2020. The competition moves to different cities to include representation from each region on a rotating basis. We often shipped cheeses across the country to West Coast cities using insulated coolers, ice packs, and the most expensive priority overnight service promising early delivery. It was hard to stomach the idea that the cheeses we carefully curated and coveted for months would eventually be sitting on a loading dock in the sun waiting for a forklift operator at a convention center, but it was all part of the competition process. We became accustomed to this ritual and won many awards.

We never actually attended an ACS conference and competition until the event came to Pittsburgh in the summer of 2018. Catherine had a brace on her knee for torn ligaments and a stress fracture, but that did not slow down the process. Al lamented that our cheeses were not as good as he hoped for various reasons throughout the five-hour car trip. We shipped the cheeses ahead of time per the rules and arrived at our hotel exhausted. Every time we left the farm, it took days to prepare the task lists and contingency

IT TAKES A VILLAGE

plans for the farm-sitters, and this trip was no different, except that we returned to the farm with five ACS medals. When they announced our name, the auditorium waved Pennsylvania flags, but before we could reach the stage to accept the medal, a new announcement of yet another medal for Yellow Springs Farm interrupted the cheers. It was a personal best showing for the farm, and right in our home state of Pennsylvania. Our cheesemaking colleagues and many partners from Whole Foods Market were in the audience. Thank you, Deanna, Adam, Mike, Dana, and all our regional colleagues. After the ceremony, there were receptions and celebrations, but we sought quiet and solitude to let everything sink in. We found a modest wood-fired pizza place with outdoor tables, ordered a bottle of Italian red wine, and gradually decompressed. We needed time to process not just the awards, but the months and years that led up to this. It was a humbling evening. More than fist-pumping, we felt the urge to reach out with gratitude and thank the many people--and goats-- who had contributed to the farm since we made our first cheeses in 2009.

Victory Brewing was among the first local collaborators. The beer originated in Downingtown, Pennsylvania, just fifteen minutes from the farm. Their brand was already filling regional headlines and winning national accolades when we began cheesemaking. When Victory put Yellow Spring Farm cheese on their pub menu, we felt thrilled to be in the company of winners. We made a custom, limited batch of cheese with truffles for Victory's fall tasting menu. Later, we returned the handshake, and began washing an aged goat cheese tomme with Victory's Storm King and Hop

Devil beers. The rind glowed with its warm amber color with yellow highlights. We called it Yellow Brick Road. We loved Elton John's song "Goodbye Yellow Brick Road" and we lived on Yellow Springs Road. Synchronicity rules the world. Matt, Diane, Lauren, and so many Victory friends helped organize many "Happy Hour" guided cheese tasting events. We loved creating these food-centric programs and filling the room with folks who wanted to explore unique cheese and beer pairings. Victory Brewing was a "rising star" regionally and nationally, and they generously shared the spotlight with us.

The collaboration reached beyond cheese and pairings dinners. A daughter of Victory Brewing's founder Bill Covaleski volunteered as a farm intern while she contemplated becoming a physician. Rest assured, birthing season offered her a sampling of blood, body fluids, painful loss, and small miracles. When we look back at the farm interns, it seems there were around 100 young people who spent at least a few weeks lending a hand. They were, and are, an unintentional farm crop. Many have since gone on to careers in farming, healthcare, public policy, government, food service, landscape design, and education. The seeds were sown; they germinated and continue to grow.

We met Terry Brett as customers of Kimberton Whole Foods. This independent, family-run natural foods store is located about six miles from the farm. It is the flagship store for the business which has admirably grown to include six more regional store locations. It has no connections to the national Whole Foods Market. The KWF store is a meeting place; it is a town center for those interested in organic foods, nutrition, homeopathy, and sustainability. They care about farms and farmers first, and are merchants second. We attended a casual dinner to celebrate 25 years of KWF, and still remember the emotional stories offered by the dinner guests in homage to Terry and Pat Brett as people, friends, and business owners. It was a pleasure to be there and see how food and farms can combine to make friendships. KWF was committed to supporting local farmers and Terry carried our cheeses and yogurts from the beginning, until we announced our dairy production was ending. They promoted our holiday cheese boxes and farm events, while we celebrated each of their new store grand openings. Together, we created an exclusive KWF cheese "Kimber-Tomme" to enhance our collaboration.

Our Yellow Springs Farm - Whole Foods Market partnership started in 2014. The store name happened to overlap with that of Kimberton Whole Foods, but the business culture and history were very different. After a popular natural foods supermarket in Devon, PA (about nine miles from the farm) named "Fresh Fields" was purchased, it became part of the national chain of Whole Foods Markets stores. We began selling cheese to the Whole Foods stores before the grocer later became part of Amazon in 2017. Building this relationship was a long process, but well worth the effort. Never was an invoice paid even one day late. They became our number one customer. It took a lot of effort to promote our cheeses throughout their 50 stores in the Mid-Atlantic States, but we could not have asked for a better partner. We did countless in-store demos with cheese samples and attended local producers' meetings and conferences of all kinds. They even held one corporate event at the base-ball stadium in Washington, D.C. As our relationship with Whole Foods Market grew, we built friendships within the management team, in-store cheesemongers, and so many associates. Mutual respect was always present in discussing big and small matters, even though the dynamic might have instead easily become a story symbolic of David and Goliath.

Whole Foods frequently opened new stores in the mid-Atlantic area. One location in particular, Exton, Pennsylvania, was exciting to us, since that was right in our backyard. Whole Foods is a big believer in featuring local suppliers to their customers, which is among their key differentiators versus more institutionally oriented grocery stores. We drove through Exton often and watched the progression as the new store came to completion. Representatives of Whole Foods did not share many details about the opening, other than that it would happen soon.

Mike, a regional buyer at Whole Foods, reached out to ask us to make a custom cheese to celebrate this new partnership so close to the farm, but it was only six weeks before the opening. Embracing our "love for local," we decided to make it a mushroom goat milk cheese. Chester County, the home of Kennett Square, PA, is the Mushroom Capital, of the World so we wanted to combine a local ingredient with our milk to make a hyper-local cheese. Mike came up with the name "Kennett," inspired by Kennett Square, PA, and the large mushroom festival held there each year. There is an unmistakable odor of decaying compost that announces your entry into the town at any time of the day, and in all seasons. Some find the odor unpleasant, but we feel it is a true expression of terroir. We loved that the milk and mushroom both came from the same climate, soils, and weather patterns in Chester County. From there, the creative process started.

When creating a new cheese, it could take months to develop a good recipe and test multiple batches. Six weeks was a challenge. To meet the opening day launch, we had to produce a cheese that could age quickly. Given the time constraints, we had to make multiple batches in parallel to see which cheese could make the cut. Some variations were different shapes or sizes. Some were cooked longer, pressed to release more moisture, or ripened differently. We had never made cheese with mushrooms before, so deciding on a supplier, the type of mushroom—fresh or dried—needed to happen quickly. We found an excellent source for dried mushrooms in Kennett Square. Now which cheese to make? We produced a cheese called Bliss. It was a Camembert-style cheese that had hints of mushroom and became soft and gooey when ripe. This cheese usually ages within 10 to 14 days and it is ripe within 21 to 30 days. Since we would modify an existing recipe, we could start making Kennett cheese and try three different variations and pick the best batch for Whole Foods Market customers.

We made the three batches, and then it was a waiting game. The first batch was not good enough. The second batch was slightly better, but not the preferred cheese we were hoping to provide for the opening. Whole Foods placed their opening store order for our other cheeses, and we delivered them. They asked us how our new cheese was coming along, and we said it was not ready yet, but that we would know soon. The Exton store opening was in a few days. We anxiously waited and after several days, we tasted the third batch. Hooray! It had the profile we were looking for. With no time to spare, we packaged and labeled the cheese and delivered it for the opening. The refinement of Kennett continued in the coming months. We knew with time, we could make it even better, and we did. Kennett launched in many more of the Whole Foods stores in the mid-Atlantic and became a successful collaboration with an important community partner.

Some community collaborations started out more tentatively. When the Chester County Agriculture Council held a panel discussion about local food as a source of economic development, Catherine was in the audience. Each panelist spoke and praised the virtues and romance of loving local, buying local, and eating local. Catherine had heard these lines before and was a bit tired of the words that hung in the air but were not regularly backed by actions. She introduced herself to panelist Chef Patrick Feury after the event. Patrick was a James Beard award winner and had been featured on Food Network TV, to name just a few accolades. Catherine mentioned that she had been to Patrick's Nectar restaurant in nearby Berwyn, Pennsylvania, but that he had never been to Yellow Springs Farm.

It was hardly a subtle hint, and within a week or so, the tides changed, and Patrick came to the farm for the first of many times. Among the constellation of community farm supporters, Patrick was a shining star. The Nectar menu featured Yellow Springs Farm cheeses in the appetizers, main courses, and desserts for years. There were unique, creative pairings for goat cheeses which were especially unexpected for a restaurant with an Asian fusion theme. The ripe Black Diamond cheese even became a savory ice cream variety on the dessert list.

Nectar had a great PR team, and Patrick willingly shared the spotlight. He included Yellow Springs Farm in many photo shoots, newscast interviews, and exclusive restaurant wine dinners. It was a convivial business relationship, but also a sincere human connection. Patrick rarely sent a staff member to pick up his cheese order at the farm. Instead, he came, and was often ready to share a hot samosa he picked up on the way over. He wanted to see the newest kids born and discuss pickling techniques for summer vegetables. We compared tasting notes on trendy gins, commiserated on the plight of small business, and shared inspiration for possibilities ahead. Visits were punctuated with a hug and kind words.

When Patrick was a featured chef at the James Beard House dinner in Manhattan, he used it as a platform to feature Chester County cheeses. It was a fun, festive, and fancy evening for us and two other cheesemakers to enjoy the party in NYC. It was clearly a ton of hard work for the culinary team, but magical to see so many top chefs dressed in their finest white (some black) toques and aprons. They seemed to dance about the open kitchen amidst glistening stainless steel and steaming pans bathed in other-worldly, appetizing aromas. This sensory memory could almost compare to opera or ballet, but it was even richer because we actually tasted everything and enjoyed the meal together. There were countless passed hors d'oeuvres, many that included our cheeses, while the chefs prepped the entrée. We don't remember any specifics about what we ate after being seated, and perhaps that is an important point. The adrenaline, chatter, handshaking, and networking kept the evening rolling so quickly, it was as if we were running on fast-forward for hours.

The event ran until well after 11 p.m. It was snowing when we walked outside, and no taxis were around. We walked back to our hotel near Penn Station in the quiet whisper of Manhattan aglow in winter wonderland. It was a beautiful and emotional way to wrap up the evening, even though we had no hats, scarves, or gloves, and the wrong footwear for snow. Given the prices of New York hotels, our modest room was tiny. The shower was so narrow we thawed the left limbs first, and then turned to bathe the right arm, leg, and foot in hot water. Gradually, we stopped shivering and went to bed, but not for long. The fire alarm sounded just after 1 a.m. It was not a drill. The smell of smoke filled the room and we heard lots of commotion in the halls. We dressed quickly; pulling on the wet shoes we had worn earlier, and headed down about six floors on the exterior metal fire stairs. No one was allowed to re-enter the building for a long time as first responders evaluated the emergency, and so we again were outside on a memorable cold, snowy night. We became hoarse and caught colds, but would not have missed it!

The following day was Act 2 of the NYC adventure. We received an invitation to Blind Tiger, a craft beer bar. It was a Sunday, so the crowd grew slowly beginning at noon. By 2 p.m., we could not hear ourselves talk or think. The place was packed with spirited beer lovers gathered for a taste-off: beers and cheese of Pennsylvania vs. beers and cheeses of Vermont. The votes were tallied—although louder fans might have stuffed the ballot box—but needless to say, Pennsylvania was victorious.

The first time we collaborated with Nectar for a special event dinner, it centered on featuring our farm's goat meat. When cow dairy farms have too many male calves, farmers often process them for veal meat. Rose veal is in high demand among consumers and commands high prices. In contrast, goat meat is not typically part of American menus. Male goats have a reputation for smelling bad, butting heads, wreaking havoc by eating anything they see, and jumping out of all enclosures. Caribbean traditions embrace stewed or curried goat. Some Mediterranean cultures look for roasted goat on a spit, or young goat roasted for Easter meals.

Chef Patrick leaped at the opportunity to purchase meat processed from young male goats raised at the farm. He embraced the ethos of wasting nothing. Annual birthing is an essential part of milking does for cheesemaking. About half the goat kids born will be male, but farms need only a few male goats for breeding. He honored the animals and the circle of life with a menu we still remember vividly. It started with the farm's artisanal cheeses, and then agnolotti pasta stuffed with slow-cooked shredded goat meat and mushroom, and next were the delicately seasoned chops. They were smaller than lollipop lamb chops, but the meat was flavorful and tender. We never enjoyed eating goat meat this much before, or since, this memorable dinner.

Catherine turned 50 years old while working on the farm. The milestone birthday was on a weeknight after a task-filled farm workday. In keeping with our food-centric way of life, a very special dinner was the best way to celebrate. Patrick prepared a special tasting menu at Nectar restaurant for the occasion. We had been to Nectar at least a dozen times for working dinners, where we were the presenters educating the guests about cheese, sharing samples and pairings with winemakers and brewery reps, and discussing recipe subtleties. But on this evening, it was a table for two, just a romantic date night for Al and Catherine. Each course included Catherine's favorite things. Patrick had listened carefully to casual conversations at farm cheese pickup stops and wowed her with dishes tailored especially to her palate. Dinner included fennel, saffron, and lentils, veggies pickled in craft gin, seasonal fruits, shellfish, pastured meats, duck breast, and artful plating. We paid for the meal, but the gift was priceless.

Patrick passed away in February 2022, a few months after we sold Yellow Springs Farm. It was a so sad to learn of his untimely death. The farm would never have been the same without Patrick championing our work and sharing our journey. He was a leader, a friend, and an ombudsman for many farm-fresh food producers. Cooking was the only way we knew to celebrate Patrick's life. We had a jar of fresh peaches a friend canned in Pennsylvania and gifted us before we moved to Maryland. Catherine is an artist and sees colors in all her thoughts and emotions. Without thinking, she made a quick cobbler with the peaches that shared the same hue of orange we knew in the upholstery, interior décor, and graphics at Nectar Restaurant. The house smelled good, and the aroma soothed our grief. Eating provides comfort, closeness, and helps process memories.

Our path forward from focusing the art of cheesemaking in early years, to later embracing the science of cheesemaking in scaling up production and meeting larger demand was hardly a straight line. Ultimately, even as the business grew, we circled back again to emphasize the creative work of recipes development. This allowed us to return our focus to local food and community. It was so inspiring to have neighbors helping gather black walnut fruits for our Nocino liqueur, and Whole Foods team members cheering for our cheeses at state and national competitions. Chefs, like Patrick, Greg, Adam, Julia, and others, were instrumental in encouraging us to unapologetically return to our roots as cheesemakers and lovers of everything local despite the market pressures to scale up production towards a process more like manufacturing than art. Sometimes you have to start at the end to get back to the beginning.

CHAPTER 9:

PORTUGAL–A PLACE
OF INSPIRATION

As farm owners, we budgeted about seven to ten days for cheese journeys abroad once every few years. We used this time to search for inspiration, and for recovery from the mental and physical toil of farming. We mainly scheduled cheese travel for the winter months—November or January—when the goats produced less milk and the plant nursery was dormant. Seeking to embrace our Mediterranean cheese palate, we wanted to travel beyond Italy and France. Winter weather made it clear we would postpone an otherwise tempting cheese trip to Greece. Spain was an option, but it is a big country with significant historical and cultural contrasts in each region. We decided it would be great to explore Spain's culinary traditions, but better if or when we had at least three weeks to travel from Madrid to Andalusia to Catalonia, and everywhere in between. We could not afford a longer itinerary while Yellow Springs Farm was operating.

Because of our shorter time frame, we needed to select a smaller country with a temperate climate; we chose Portugal for our 2011 cheese destination. *Tur Aventur*, a mother/daughter team focused on travelers who wanted to cycle, kayak, and go off-roading in Jeeps, helped organize our itinerary. Our sense of adventure in Portuguese culinary and farming traditions offered an unusual but interesting direction of exploration for this family travel planner.

Serra da Estrela is the highest elevation point in Portugal. Even on this region's rocky slopes and ledges, shepherds make this eponymous traditional Portuguese cheese variety with raw sheep milk. Because we were familiar with Serra da Estrela cheese, we had misconstrued images of mountain goats in our minds when we first planned to visit Portugal. Both goats and steep slopes were worlds away from the pastoral plains in the Alentejo region, extending south from Lisbon to the coastal areas of the Algarve that we actually visited. The trip began in Lisbon, headed west to Sintra, and then south throughout the Alentejo region. The scenery suggested there were likely more sheep than people in many municipalities. Back to scenery for a moment—it was all-consuming, otherworldly, with sensual stimulation in every direction. The meeting of ragged cliffs over ocean vistas with infinite sightlines punctuated with ancient and medieval stone structures is unforgettable. We loved learning about the native oak trees (Quercus suber), from which people harvest cork once every 9 to 10 years. We learned that the cheese existed mostly as a servant to wine, so it would be imperative to taste and understand Portuguese wines to learn about its cheeses. Maybe this wine tasting itinerary sweetened the visual memories of the scenery, but nonetheless, it was a spectacular adventure.

We were told in advance that sheep outnumbered goats by a large number in Portugal. We assumed it was a tradition born from practical use for the sheep fiber that offered a chance for spinning wool yarns, and making warm winter blankets, coats, and boots with sheepskins. Another couple visiting the area joined us for breakfast at an agritourism home after we had visited a few cheesemakers raising sheep. They were Portuguese physicians who spoke fluent English. It was cordial, yet casual, with a warm cultural exchange of personal tidbits. Then, as the topic turned to goats, they told us about the Fever of Malta. Humans contracted this dreadful disease in the 1980s through the consumption of unpasteurized goat milk and goat cheese. Many Portuguese were hospitalized, and some died. It was still a recent memory for many citizens, so goat milk was seen as dangerous food akin to wild foraged mushrooms with questionable identity, or raw seafood of indeterminate origin.

We politely listened with measured emotional reaction but took careful mental notes. Upon return to Pennsylvania, we immediately researched this mysterious zoonotic disease to learn it is what Americans call brucellosis. Caused by bacterial infection, it is treatable with antibiotics if identified and diagnosed quickly. Our trusted veterinarian, Missi, patiently assured us that our goats were very unlikely to have brucellosis, as they would catch it from contact with sick cows, and we had no cows. Besides, Pennsylvania dairy was determined to be a brucellosis-free state through testing, culling, and vaccination, according to its Department of Agriculture. We were then only making pasteurized milk cheeses at Yellow Springs Farm, but the fear of this plague-like path to disease, destruction, and maybe death set us up for serious introspection and self-regulation. It reaffirmed that we wanted to make the very best cheese possible, with spotless production conditions, and careful herd health management.

As we sampled more Portuguese farmstead cheeses at farms and open-air markets, we saw they were either stark white or covered with a smoked paprika' s mottled red and orange pigments. We shared photos of our cave-aged, natural rind Pennsylvania cheeses with Portuguese producers, only to see they were politely aghast. They marveled that customers would dare sample or purchase cheeses that showed signs of dirt—or what we proudly called earthy terroir—generated from the molds and yeasts from our 19th century bank barn and cheese-aging cave.

As we entered more producers' facilities, we observed procedures where they handwashed each piece of cheese with salt water and/or vinegar to eliminate any natural surface molds. We know molds and yeasts break down milk proteins and fats, and this process develops cheese flavors. The Portuguese goat cheeses were instead going to get flavor from surface coatings of paprika or from the all-important wine pairings.

Pastured cows, especially dairy breeds such as Guernsey

and Jersey, and to a lesser extent goats and sheep, often have a creamy or even a yellow-orange tinge of color in fresh milk. This color comes from the carotene in the digested plants the animals ate outdoors. Carotene in carrots makes them orange, and we associate it with fresh, healthy food. But a tinge of orange in cheese was likely to trigger the fears of dirt, disease, and risks of death associated with past outbreaks of foodborne illnesses. We saw producers add chlorophyll to the vat as they made cheese so that this natural pigment would absorb any residual carotene and assure that the cheese came forth seemingly whiter than white exists in nature. To our eyes, it reminded us of bleached linens in a hospital setting where sanitary needs trumped any romance for flavor, aroma, or palate-pleasing character.

Loving the landscape, we willingly traveled many kilometers on country roads from village to town to knoll to coast and back again. Sheep were everywhere, along with a few cows, but we still never saw a grazing goat. A wonderful family of cheesemakers hosted us, with four generations of women who delegated tasks in the production room according to each person's abilities and aptitudes. The great-grandmother sat prominently at the head of the room, where her eyes ruled. Her sense of quality control was ageless, as she wrapped finished products in carefully folded papers with pleated corners and marketing labels. We finally asked the burning question, "You have goat cheese, but no goats. Where are the goats?" They explained that they kept the goats at a separate farm, several kilometers away. To avoid risks of cross-contamination, they kept the goats segregated and isolated from other animals and people. "Blame it on the goats," was the refrain. Call it bias, discrimination, or perhaps a vestige of difficult experience. The artisanal dairy tradition of the Alentejo did not value goats.

Although Portugal joined the European Union in 1986, it seems many of the agricultural regulations were trickling down more slowly and started to affect producers materially after 2000. We saw two approaches to meet the new regulatory environment for cheesemakers.

In one case, a woman proprietor of a very small cheese plant used government loans and grants to purchase state-of-the-art equipment. She called her place a cheese factory, and that it was. There was no sign of animals or plants. Milk arrived by truck and cheese left by truck, going out doors on the other side of the building. In between, she used an impressive ultra-pasteurization system that moved milk through in 30 seconds. To save floor space, the system used a vertically formatted serpentine of stainless-steel tubes and pipes. Coming from an electric pasteurizer with a hot water jacket that took a few hours to heat each morning, and several hours to process each batch of milk, we were amazed at the efficiency and productivity we witnessed. Just two or three people were working, but the equipment and mechanization made lots of cheese six days per week and ran long hours. Scalable production made for a profitable business. Sure, there were loans to pay, but machines did not require vacation days, sick days, or other obligations from their employer. Labor shortages continue to be a limiting resource for small-farm productivity worldwide and here was a 21st century solution in action.

On another day we visited a sprawling farm with various

low-rise cinder block buildings freshened with white stucco and trimmed with eye-catching azure blue. Why blue, we asked? Portuguese farmers and other many cultures traditionally used this color to repel evil spirts or attract benevolent spirts. It was a good luck charm of sorts. In contrast to the science-based business we saw earlier, this place was defined by more organic, emotional energies. On arrival, a few large, loafing dogs greeted us. They were not threatening, but nonetheless announced our arrival. When I asked about the breed, the owners explained they were Alentejo hounds—farm dogs that have been raised in those parts for generations. They come with the land and are part of the family traditions. When we adopted Roscoe, a husky/bloodhound mix, from a dog rescue several years later, we selected him with the emotional memory of the Alentejo hounds' floppy ears, lumbering gait, observant eyes, and laidback vibes.

During almost four decades of the Salazar dictatorship (1932-68) the Portuguese government seized land belonging to bourgeois families. The idea was to return these farms to the workers who could ideally become self-sufficient and support themselves in the rural lifestyle. Most of the owners retreated to the bigger cities, where they practiced professions or operated small shops. After 1974, as a new government formed, changes in policy allowed the former families to claim their agricultural holdings once again. The younger generations, tired of city life, came to the land seeking to revive the traditions of their grandparents and great-grandparents. They were mostly well-educated and had access to capital. Many were already entrepreneurs and

found the farms an enticing lifestyle. They brought inno-
vative business plans for producing diverse value-added
products such as olive oil, herbal products, handmade pot-
tery, and fiber arts. Coupling these items with hospitality
of farm stays in bed-and-breakfast style accommodations,
they created a unique experience. They had sheep dotting
the landscape, and they had the barns, milking parlors,
and creamery buildings on site, but they were not making
cheese. Why?

Unlike prior generations, the 21st century farming generations were subject to much more regulation. The inspections, permits, and physical plant improvements needed
to produce artisanal cheese made it economically impossible to justify the investments needed to satisfy EU regulations. Not deterred, but simply detoured, the families began
changing cheesemaking focus to an agritourism offering.
They offered participatory classes, where visitors of all ages
could see the process, get their hands involved, and even
taste the cheeses made by the group. Adults and children
came from cities for a day trip, or from foreign countries
for a weekend farm stay. Cheesemaking differentiated the
offering on countryside properties, called Quintas, and provided a revenue source greater than the business opportunity of cheese production for wholesale or retail markets.
Farms were happy to leave cheese production to the cheese
factories, focusing instead on satisfying the desire for the
romance of cheese.

Upon returning to Pennsylvania, we embraced this inspiration. We began offering cheesemaking classes on select Sundays, and for a few special guest groups at Yellow

Springs Farm. The classes sold out weeks in advance. Our cheeses were popular, but people loved our classes like they love rock stars. Typically, we had around five to six attendees in our classes. The goal was to give them a real farm experience. Their first task was to milk a goat. Since most people purchase their milk from a grocery store, few people realize where their milk comes from. Milking a goat is a tactile experience. It is a moment when you come face-to-face with a soul that happens to have floppy ears, four legs, and an udder. All our goats have names, and Rena typically steals the show because she consistently behaves well on the milk stand and is welcoming to new people. Imagine never having milked a goat before, and now you find yourself bending down, touching an udder full of milk, cleaning the teats, and then gently squeezing the teats to release the milk into the waiting bucket. It can be both an exhilarating and fearful experience for people, but one that leaves an indelible memory. At that moment, they realize how precious this gift is from the goat and the special care that is required to curate this substance into a cheese that reflects the terroir of the farm. We would take the milk from the morning's milking and make goat milk mozzarella cheese in the class. Making this particular cheese gave the attendees an opportunity to cut the curd in the vat, stir the curd and, when ready, feel the curd when the mozzarella curds were placed in the cheese molds to drain. For most visitors, this was the first time they ever felt warm mozzarella curd with their gloved hands. They learned the importance of food safety, gently placing the curds in the mold so as not to lose the proteins, fats, and minerals that give the mozzarella the terroir of the farm. Most importantly, they had the opportunity to taste some of the curds while they were still warm and squishy and before being drained. After the cheeses drained for thirty minutes or so, we would take the cheeses out of the molds, add a small amount of salt, and serve them with tomatoes grown in our garden for a real farm-to-table experience. Sometimes our cheesemaker students even brought wine to share to enhance the festivities and round out the day.

Even though the cheesemaking classes were fun, profitable, and allowed for community connection, they were not sustainable. Like the Portuguese model, we saw that cheese classes were a significant part of a diversified value-added farm business, but not a plausible stand-alone offering. Besides, given the 24/7 nature of our farm life, it was difficult to add this to the multitude of chores that we had to accomplish. Sunday was never a day off for us. We milked the goats, turned the cheeses, and prepped orders. However, with one less extracurricular activity of conducting a class, we were relieved to have slower days on Sundays.

To sustain a farm business for 20 years, there is a requirement for balance and seasonality of efforts, so the tenure of teaching cheesemaking ended as our cheese production grew. The call for diverse revenue streams and need for improving profit margins were exactly the same as the challenges faced by our Portuguese dairy peers. Our trip to Portugal was a time for rest and relaxation, but also a source of inspiration. We thrived on more cultural exchanges like this, since we realized how important it was to break out of our day-to-day routine to be rejuvenated to continue our farm journey.

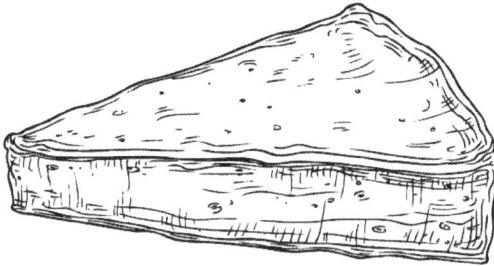

CHAPTER 10:

WHAT IS AMERICAN CHEESE?

W e had eaten plenty of those yellow/oranges slices of plasticized product wrapped in clear film that we called American cheese—grilled cheese, sandwich cheese, burger toppings. These were staples in the family kitchen, school cafeteria, and even some restaurants when we were younger. After Yellow Springs Farm goat cheese won American Cheese Society awards for five consecutive years (2009-2013), we were more interested in parsing the adjective and noun pairing. Exactly how do we define American cheese in the 21st century?

Europeans arrived in Pennsylvania in the 1600s. With the Declaration of Independence and later the Constitution signed in Philadelphia, it was easy to believe America started there in some way, shape, or form. We casually use the word "America" interchangeably with the United States of America and the USA. But that does not even consider the differences between North America, including Canada and Mexico, plus South America and Central America. Each of those is America, too.

In January 2014, we traveled to Uruguay. It was our first cheese experience in South America. We sought to understand if the French and Italian influences on traditions carried forward by European immigrants in cheesemaking looked and tasted like it did in the USA, and specifically in Pennsylvania. Our hosts repeatedly said that we were arriving in America, as we explored Montevideo, and the countryside beyond. America it was—simply not the United States of America.

Our travels in Europe reaffirmed our supposition that wine and cheese travel the same cultural and culinary roads. Areas that produce great wine almost always have a few outstanding cheeses. These pairings work well together. Both fermented agricultural products are born of the same soil, climate, and culture, so of course they are a splendid match.

With this background, we first started our query about South America by looking into a trip to Argentina— Mendoza and its Uco Valley were rapidly growing wine regions with increasing lists of accolades. But here, surprisingly, the wine and cheese patterns dissolved, unlike in Europe. We talked with travel specialists about Argentina. Some were uninspired, but others paused and asked more pointedly, "Why are you going to Argentina?" We explained our passion for hyperlocal food, artisanal production, unique microclimate, and terroir. We felt the lens of South

American would show us new discoveries of these near and dear passions, and perhaps inspire new cheese varieties or better cheesemaking processes.

The pivotal response from a Texan travel professional was profound. He told us Argentina was chock full of cheese factories, mega farms, national brands, and commercially distributed cheese. He felt there was little interest in or demand for small-batch cheese in Argentina, but that Uruguay would be a much better destination for our travel goals.

Despite our doubts, we were intrigued. We listened on the speakerphone, as we began typing to see if we could spell Uruguay correctly, and then locate it on a map. We asked for a proposed itinerary and waited a few days. The idea was growing more interesting, so we made air reservations for flights to Montevideo via Miami and marked the calendar. We knew it would be summer in February in South America, and perhaps the promise of mild weather was enough for now.

Rosario was our guide in Uruguay. She was loquacious, energetic, and congenial. She was married to a butcher. This was our first clue about the food culture in Uruguay. She told us on the first day that the people of Uruguay have three loves—soccer (football), beef steaks, and Tannat (a local red wine). No, cheese did not make the list of the top three, or the top five, but perhaps sneaked into the top 10 food trends at some fine restaurants and international hotels.

It felt like Uruguayan goat cheese was experiencing an early trending renaissance, perhaps analogous to goat cheese's rising popularity in the early 1980s in the USA. There were a few producers making small batches of cheese, but there was not much goat cheese, or cheese of any kind, in supermarkets. When we walked in small towns, there were cafes, butcher shops, and wine shops, but not a single shop dedicated to cheese. Cheese was in the back of the winery, or in a small case, beyond the meat. It was an afterthought, or perhaps a garnish—a curiosity, at best.

Rosario drove us outside the capital of Montevideo through expansive marshlands. Rice is an important crop there. Cattle grazed without fences and gave new meaning to the terms "free-range" and "grass-fed." Locals quipped that investing in a bull calf assured you could double your money in two years by processing the animal for meat, but no bank in Uruguay could offer such competitive returns or financial security. Despite being often compared to Swiss banks in Europe for their stability and attraction of foreign assets, the banks in Uruguay were no match for the grasslands that feed young cattle.

When we visited our first goat dairy, we met a young man with a wool beret eager to share his herd and his cheese kitchen. We compared notes on animal nutrition, parasite management, breeding cycles, and milk quality. He spoke English well, was an excellent herdsman, but still learning to be a cheesemaker. He had modest means, rudimentary equipment, and aging buildings, but great pride and ambition. We were confident he would grow his business.

Our next cheesemaker introduction was at a winery. It was not a farm, and there were no animals in sight. Here the milk was purchased from farmers who delivered product. Our cheesemaker hostess lived on the property. She was a divorced woman with an admirable entrepreneurial spirit.

She served us beautiful, layered confections that resembled birthday cakes. The cheesemaker decorated the cheese rounds with edible flowers, and each layer showcased the unique color and texture of a chosen cheese variety. She thought of cheese as a creative medium to charm, enthrall, and appease sensual attractions. Cheese here was trendy, sexy even. It was all about how it looks—taste, recipes, and wine pairings were not the priority.

We spent time at the coast on picturesque beaches with dazzling purple sunsets stretching as far as the eye could see, almost wrapping towards the moon and stars. We saw art galleries featuring local painters and sculptors. There was high design everywhere in the architecture of single-family vacation homes and urban commercial buildings. Color and detailed attention to form, texture, and design were prominent in textiles, housewares, and just countless details of everyday life. Uruguay had an insatiable appetite for aesthetics, especially in the wealthy, coastal area near Punta del Este.

Touring the countryside, we enjoyed meals of beef cooked over an open fire, served with Tannat. It was easy to know why locals loved this tradition. We walked through vineyards and saw the grapes hanging in the blazing sun. The more we got to know Uruguay, the more we knew artisanal cheese did not live here. If it was not beef, the dinner featured shellfish. The clams were particularly beautiful and flavorful. The shells were smaller and glossy with translucent swirling, and black with white patterning. We learned in Italy the stern cardinal rule that one should never, ever, have cheese with shellfish. If clams, shrimp or mussels were on the menu, it was a cinch cheese would be left out.

The question was the answer. The European traditions of French and Italian culture arrived in Uruguay, but they were not recognizable in the cheese culture. Uruguay did not have the threaded chains of microclimates and local appellations we found in the Alps, the Loire Valley, and Piedmont, Italy. In Italy, and France, and even in Portugal to a degree, wine and cheese pairings were as inseparable at the dining table as they were at the soil, in the vineyard, and the farmyard. There was little doubt that Uruguay loves soccer; beef, and Tannat much more than cheese.

CHAPTER 11:
ENOUGH IS ENOUGH

We enjoyed our cheese travels immensely, but they were brief interludes in the everyday reality of running a small farm business. The travel days were especially precious for rest and recharge, because otherwise, if we were at the farm, it was a workday. When did Monday start? Not when the sun rises, the dog barks, or when the goats come up for milking. Monday started overnight on Sunday while we are sleeping and purchase orders from Whole Foods Market stores, our distributors, local specialty shops, and chefs arrived via email. This was the start of a weekly ritual involving counting, packaging, and planning delivery routes.

While tallying the orders and the inventory, including just-in time-cheese that would ripen until the hour of delivery, it was always a matter of quick cash flow arithmetic. When would the order payments be received? When was the next payroll date, and what were the autopay dates coming for credit cards, insurance, and utilities? It was a Match. com moment in trying to pair production, fulfillment, and payments, while unrelentingly testing cheese quality and food safety. We never forgot that including love and delight with each box made the cheese taste better. Passion made the difference. Hourly, we surely earned less than minimum wage, but we chose a lifestyle, a connection to community, and the priceless satisfaction of being makers. Owners' eyes touched every order, and for that we were proud, but often exhausted, and constantly watching the spreadsheets, accounting, and business finances. We committed ourselves to sustainable agriculture; that included sustaining healthy people, goats, plants, planet, and profit. It sounds simple, but running Yellow Springs Farm was the hardest job we ever had.

It is a sort of reverse engineering process that figured out how to get the requested cheese to the correct location by Thursday morning of the same week, all while keeping the products in temperature-sensitive compliance. Thursday delivery targets left us one day of grace on Friday in case things went wrong, such as ice storms, power outages, flat tires, printer errors, floods, and eventualities too numerous to name. Sometimes we used insulated shipping cartons with ice packs sent overnight by UPS or FedEx, but more often we delivered cheese in coolers packed with ice packs within a refrigerated van. Almost everyone at the farm drove the delivery route at least a few times. It was a bit like being an air traffic controller to figure out the logistics of packing, invoicing, staffing, shipping, and driving. The order of stops, the traffic delays, vehicle fuel and maintenance needs, plus the schedule of mandatory receiving hours specific to each store location. We imagined what the takeoff and landing charts at a major airport look like in real time, and realized this was a good paradigm for us to emulate. We organized countless moving parts, each with a precise destination, and tried to avoid the risk of a domino effect of failures, or even a crash, if even one component breached the farm management system.

Order fulfillment was stressful. During the holiday season, we shipped cheese sampler gift boxes, too. While others decorated for Christmas, we called in more elves to pack boxes, label cheese, and make gift cards. Our reverse dating calendars helped us identify the optimal ship dates for every destination zip code. We divided the shipments into two groups: ground and that needing 1- or 2-day air service.

We tared the empty containers and standardized the packaging to estimate the final weights and dimensions accurately for shipping labels. Each morning we dreaded the flash of "Exception" in the UPS tracking messages. We heard all the perils and causes for delays, including bad weather, labor shortage, equipment breakdown, and road closures. Our temperature-sensitive cheeses were unforgiving. They needed to arrive on time. Some days went smoothly, but others ended in us reciting the "Serenity Prayer" for patience to accept the things we could not change. Red wine helped ease the tension. There was always hope for a better tomorrow.

The cheese business was going gangbusters in 2015. The

sales were paying the bills, even burdened with equipment loans, insurance, and payroll. Cloud Nine sold out most weeks, even before we made it. We ripened Black Diamond cheese with vegetable ash. The white surface with black highlights reminded us of a designer gown for a formal cocktail party. Black Diamond was the "Belle of the Ball" and her presence was in demand, especially at fine restaurants. Our cheeses retailed for about $35 per pound. It was top price to begin with, so we were surprised when a customer sent a photo of Black Diamond selling for $49 per lb. in a Manhattan specialty shop. They too sold out.

All we could do was make more cheese. We resisted the idea of running two shifts at the farm to make cheese twice a day. The stress and lifestyle change would have been unappealing, or worse. We understood the downstream problems of ripening, aging, refrigeration, packaging, and shipping would still create bottlenecks, even if we somehow had more milk and made more cheese. Once we upgraded to a 100-gallon pasteurizer, two iterations up from our 2009 used starter model, we felt the cheesemaking was at its maximum volume. There was no more floor space in the creamery. The bank barn had 20-inch-thick fieldstone walls. Building an addition was not possible for so many reasons.

Entrepreneurs know that a business is either growing or falling behind. There is no such thing as stasis. One year will never be the same as the prior year. You are either moving forward or slipping back. Markets are evolving, costs are rising, tastes are changing. We sought ideas for growing Yellow Springs Farm dairy. We wanted to stay afloat, swimming as a strong regional provider, striving to avoid drowning, rather than greedily expand for the sake of growing bigger with national distribution.

Goat milk yogurt seemed like a means to grow the dairy business. We hoped to service many of the same customers, but simply add another product to our delivery routes. Many customers chose goat cheese because of allergies or intolerance of cow's milk. These folks were eager to see the yogurts from the farm. Yogurt doesn't need space in the ripening and aging cave. The yogurt is ready for shipping the day after production. This cycle is good for cash flow and saves space in the creamery. We researched scaling possibilities for yogurt and found that it could move to a leased manufacturing facility off the farm as it grew. Cheesemaking required judgement and an artisanal touch; no two batches were ever the same. In contrast, yogurt was better served by seamless repetition. These products were a great match for automation and, eventually, contract manufacturing arrangements.

Many people make yogurt on the kitchen counter. Historically, yogurt was likely first created by leaving old milk at room temperature, so the cultures grew and fermented the product. The food safety and regulatory profile for selling yogurt is substantially different. At first, we assumed that making yogurt would be similar to making cheese, but then we discovered that the USDA categorized yogurt sold outside of Pennsylvania as a Grade A product, while cheese was not subject to Grade A regulation.

How could that be? We learned that the roots of this regulation come, at least in part, from trade agreements with European countries. Since the United States imports cheese that is not subject to Grade A rules, those U.S. cheesemakers supplying cheese to U.S. markets competing with the European imports were exempt from those regulations. We modified the floor plan in the creamery, adding two doors separating the goat milking parlor from the equipment wash area and moved floor drains. We created several more binders for animal health recordkeeping, product recall preparedness, HAACP plans, and invited one more farm inspection twice per year. In order to comply with the rules, we made changes to the printed labels. It costs time and money to become a Grade A dairy, but we saw it as an investment in our future. There was no tangible improvement in the quality of our operations, or the taste of our products. Grade A does not improve the typical consumer experience. It is really designed for much larger-scale manufacturing operations. Somehow Yellow Springs Farm had to squeeze into this framework because there are no exceptions if you are a small scale operation. It seemed crazy, but Yellow Springs Farm was being monitored and inspected by the same personnel who worked with Wawa, a regional brand of about 1000 convenience store locations packaging private label dairy products in Pennsylvania.

The last part of the Grade A yogurt making saga came in selecting the yogurt filling machine. It was not okay to ladle the yogurt from the vat into plastic deli cups. We had to purchase a closed system that pumped the yogurt into cups on a moving carousel and mechanically sealed each with foil. There was no human touch, and a very low risk of inadvertent product contamination. Research narrowed in on a Wisconsin manufacturer with a model to suit our needs. The price was shocking, but we obtained approval for commercial financing. We waited months for an update. Finally, we received a call indicating that the machine build was far enough along, that we could visit the facility where it was being built and obtain training on the yogurt cup filler. Al flew to Madison, Wisconsin, to discover the machine build was not as far along as we thought. It was clear that we were considered a small project compared to other fillers that they were working on. The manufacturer did have enough of the machine completed so that they were able to provide a brief demo. But, it did not turn out to be the full training that Al had hoped for. After another six weeks, we finally received a call indicating the machine was completed and final payment due. We had delayed final payment after the demo, since we did not feel they had met their goals for completion. After a bit of back and forth, we finally made the last payment and they agreed to ship the filler by ground transport. The machine arrived on a box truck with a lift gate. The courier left the machine in the driveway at the base of the ramp to our creamery, strapped and banded to a wooden pallet. It was left to us to figure out how to get this large machine into our tiny production room.

Before ordering the yogurt maker, we measured everything from door openings to ceiling height. We anticipated electrical circuit needs and receptacles. We thought we were ready, but discovered we were not. The machine measured a bit larger than the specs received before delivery. We took the metal railing off the entry ramp. Our friend Rich came with superior carpentry skills and assorted tools to disassemble the door frame. We stripped everything down to the 19th century stone and were still inches too big to fit the machine through the opening.

We invested so much worry, time, and money to get this far, we could not believe that two inches would stand in our way. An idea hatched that the diagonal dimension of the door opening would be slightly larger than the horizontal opening. Everyone remembers geometry class and knows that a straight line is the shortest distance between two points. We had to convince this bulky behemoth of a machine to turn sideways. The unlikely hero was a car tire jack. Once we got the yogurt machine close to the door, we used the jack to raise one side to make a diagonal. There were three people in back and three in front as we held our breath and miraculously guided the yogurt filler to its station in the creamery.

The stress and difficulty of the delivery day was unfortunately an omen of challenges ahead. There was nothing straightforward about the years Yellow Springs Farm made yogurt. We first tested the machine with water to learn how to set the timing and spacing so that each item filled and closed as expected. It took days of testing before we were ready to package yogurt. As the machine hummed and the pump pushed yogurt, sometimes the cups lined up with the spout. In other tries, the cups stuck together, toppled over, or filled only partially. Some foil seals burned when the temperature was too hot, and others were off center, so the container was not closed. Each yogurt packaging run was supposed to last about an hour. If you remember the *I Love Lucy* episodes when she and Ethel briefly worked in a candy making shop, you have a good idea of how many things could disrupt our workflow. Practice did not make perfect, but over time we expected problems, became more efficient, and reduced the product loss during packaging.

During our yogurt recipe development months, we worked with a consultant from Vermont. We shipped him samples for his evaluation, and he even spent a couple of days at the farm to fine-tune our efforts. After months of sales, we eventually submitted our yogurts to competitions to see how they performed against more established brands. The yogurts won awards at the American Dairy Goat Association and American Cheese Society competitions. They tasted good, but that was not enough. We ran into more obstacles. National brand yogurts have an eight-week shelf life. This is great for retail stores. Our yogurts had no preservatives, and we only dated them for six weeks. Retailers did not like this. They understandably did not want to lose money when expiration dates required price reductions or product waste. Many times, a grocery buyer handled the yogurt purchase orders, while a specialty food buyer bought the cheeses. Even one small store would typically have us work with two different people, and two different purchase orders, one for yogurt and one for cheese. Our hope for efficiency and growing the average order size to add profitability to each delivery stop diminished.

The consumer experience with yogurt is fickle. Yogurt displays have rows of choices from floor to ceiling. It is hard, even with great graphics, to capture the eyes of the shopper. There is no opportunity for informational signage or customer service staff on hand to offer samples, or to explain each yogurt's features and benefits. The yogurt aisle is Darwinian in many ways—each yogurt is competing on its own, and only a few will survive to retain the coveted shelf space. The biggest brands get the center shelves and more fronts, so they are easy to see. Small makers like Yellow Springs Farm might be placed closer to the floor or high above eye level. When stores stocked our yogurt, given the finite space, they might take only two flavors, and resist offering the customer the larger quart containers.

Soon it became clear that making yogurt was hard, and selling yogurt was even harder. Yogurt did not improve farm profitability or offer a business growth opportunity. The venture failed. It was time to be brutally honest with ourselves and decide when it was better to end yogurt making instead of battling the headwinds and detours at every turn. We were fortunate to find a buyer for the machine who was a cow's milk yogurt maker looking to increase production. We shed tears when we parted ways with the yogurt filler—tears of joy to see a difficult chapter end, and tears of grief for the personal wear and tear we now forever associate with yogurt. Goat cheese was like our first child. We are so grateful it came before yogurt and gave us a wonderful experience to build confidence and customers. If yogurt had come first, there might never have been any Yellow Springs Farm cheeses.

As that tide went out, it opened a space for incoming tides of change. Instead of dwelling on what was going wrong, we shifted our focus to what was going right and exhaled. We were wiser and happier without yogurt. We shipped the last yogurt in Fall in 2018, just before we went on a cheese journey to Sicily and Italy. Rest, relaxation, and inspiration were again in order. Our focus up to this time had always centered on creating a sustainable farm ethos. We had now learned from this experience that sustaining ourselves was equally important.

CHAPTER 12:

SICILY–GOING HOME AGAIN

In 2018, our artisanal cheesemaking business was in full swing and generating most of the farm's revenue. Morning, noon, and night, we were in the cheese room. To ensure proper cheese aging, cheese rotation, brine removal, and pasteurizer cooling, we set alarms to ring at all hours, seven days per week. We were shipping to Whole Foods stores throughout the mid-Atlantic states, managing our Cheese CSA program, attending farmers markets, plus selling to specialty restaurants and independent foods stores.

We needed a break. The 24/7 grind of operating a farm with two businesses, goat cheese and native plants, eventually got to us and the only way to get away is to, well, get away. For us, that meant leaving the country. Otherwise, we feared people would eventually find us, and then we would have to return home to deal with whatever 911 crisis the day featured.

Given that both of us have Italian heritage with ancestors from the southern Italy and Sicily, we thought a trip to Sicily sounded like a good idea. Our previous trips to Italy had been focused on the Tuscany and Piedmont regions, so a trip to Sicily hit closer to home than we have previously experienced. We were interested in warm weather, rest and relaxation, and finding Sicilian cheesemakers who would be willing to see us. It would serve as a good cultural exchange. We would learn something about cheesemaking and perhaps come home inspired and refreshed to begin anew on our journey to cheese nirvana.

We booked our stay at a small countryside hotel, Donna Coraly Country Boutique Hotel, located in the Siracusa province of Sicily. During the preparation for our trip in fall 2017, we corresponded with the innkeeper, Lucia. This property had once belonged to her aunts, uncles, and grandparents. She spent many summers at Donna Coraly as a child. We asked if she could arrange for us to meet several local cheesemakers. We explained how we wanted to better understand Sicilian cheese culture and learn how we might apply some of the art and science of Sicilian cheesemaking into our own cheese craft in Pennsylvania. Lucia was curious and enthusiastic. She said it would not be a problem, and that she would be happy to make such arrangements. We trusted her multi-generational roots in Sicily, and her worldly savoir faire. We agreed to leave the future arrangements in her hands and await the full itinerary upon our arrival months later.

Prior to our arrival at Donna Coraly, we did not fully appreciate the historical significance of this location. A historical marker on the family's land identified the spot where the Armistice of Cassibile was signed on September 3, 1943. The document confirmed the end of World War II hostilities between Italians and Anglo-Americans. When the agreement was made public days later, Germany responded by attacking Italy, freeing Mussolini from exile, and the Germans occupying many areas of the country. We know the war story's ending, as Italy formed a Resistance movement in coordination with the Allies.

We arrived in Sicily ready to relax and wind down. Much to our disappointment, there were no plans for our cheese-maker visits, despite our anticipation and excitement. "Just in time, but not early" is often the rhythm of how things go in Italy. The staff assured us they would work on the itinerary right away and get back to us. We were too tired from the trip, so had no energy to worry. We knew from previous travels to Italy that sometimes you just have to trust the Italian process and imagine what can go right. It helped that Catherine spoke fluent Italian. It did not help that our last name is "Renzi." Unapologetically, the hotel staff proclaimed, "We hate Matteo Renzi!" who was a recent Prime Minister of Italy. He bailed out the banks and favored the northern provinces, given that his political career began as mayor of Florence. We assured our hosts we had no relationship to politics and softened the mood by showing them our farm cheesemaking picture book made with Shutterfly. We were not sure how it was going to work out, but we felt that our face-to-face connection would at least be a catalyst for action. They knew we were not typical tourists because we had a farm, livestock, and food heritage. I believe they sincerely wanted to help us. After all, family and food are love in Italy.

The next day, Lucia and her assistant, Claudia, made calls and were still in the process of setting up appointments, but the situation was promising. They had our first visit planned, and they were working on scheduling several other cheesemaking visits. Things were starting to come together.

The landscape of citrus orchards, gardens, and historic stone buildings at Donna Coraly was beautiful. Cats and dogs added to the milieu, making us feel at home. Lucia had a group of dachshunds, each named after an American first lady. Notably, none was called Melania. Regina, aptly named the Queen, was in charge of greeting guests. She was a stray mutt—likely a corgi/terrier cross—but carried herself with attitude. Little did we know when we arrived that Regina would be our escort each day for morning walks. Regina was dedicated to making sure we knew the way back to the hotel. It was the myriad of such little things that made the trip special.

Breakfast was a remarkable occasion each day in Sicily. The inn had only five or six rooms, but it had a resident pastry chef. Valentina loved her craft and surprised us each morning with sweets that looked too good to eat. Of course, there was cannoli, which we are all familiar with and love. She made it with homemade shells filled with fresh ricotta from nearby farms. We also loved Cassata de Sant'Agata. The round sponge cake pastry, made with ricotta and dark chocolate, is wrapped in a sugar glaze, and topped with a candied cherry. She shaped it to resemble a human breast, with the cherry as a nipple, in homage to a martyred Sicilian patron saint. There was some synchronicity even in pastry with the theme of lactation, and our daily milking routines at the farm.

Each evening about 6 or 7 p.m., the Italians have aperitivo hour. It is a "slow food" version of cocktail hour, a time to relax and anticipate the 8 or 9 p.m. dinner hour and have a drink with some specialty appetizers and nibbles. Lucia was a gracious hostess and would mingle with the sociable group of guests most evenings. She spoke several languages, highlighted cultural commonalities, made introductions, and facilitated lively conversation, sharing her love of Sicily.

Having a historical context of the cheeses of Sicily is helpful in understanding where the traditions come from. Sicily's cheese and dairy culture dates back thousands of years. With no refrigeration, cheese was a way to preserve milk and the nutrients that come from this important food. We usually go into a store or cheese shop and think of Italian cheeses in a generic sense, without taking the time to appreciate and understand that the concept of Italian cheeses as heterogeneous. Italy is really a collection of provinces more than a unified country. Its food culture depends on differences in terroir, landscape, and microclimates. Cheese terroir can also change within a few kilometers, from farm to farm.

Our first cheese visit in Sicily was Casa Damma, run by a fourth generation cheesemaking family. The night before our visit, Claudia informed us that cheesemaking starts at 3 a.m. We wanted to learn as much as we could about Italian cheesemaking, but we also wanted to get some rest, so we asked if a 6 a.m. arrival would be okay. They agreed and said we would miss the early stages of cheesemaking, but there would be plenty of time to see mozzarella in the molds, and the ricotta in progress.

Ricotta is a word we take for granted. The literal English translation is "recooked." Most know it is the typical cheese filling in ravioli pasta. It is also used in making lasagna and cannolis. Most ricotta you see in grocery stores is made with whole milk. It is processed as a cooked cheese, but not a re-cooked cheese. What this means is when you make a typical goat, sheep, or cow milk cheese, after cooking the milk, and collecting and molding the curds, the leftover liquid whey is strained from the curds. Most of the time, the whey is disposed of or composted. If you want to make authentic ricotta, you recook the whey to 185 degrees Fahrenheit, which then causes the whey proteins to precipitate out of the liquid, resulting in ricotta cheese curds. We were told by our Sicilian hosts that goat milk whey makes the best ricotta, so that was exciting. They used the analogy of the Catholic trinity to describe the Sicilian preferred hierarchy of ricotta milks. Instead of reciting "Father, Son, and Holy Spirit," they substituted "Goat, Sheep, and Cow." It was an unforgettable scene to see the sunlight streaming through the cheese room windows intermingling with the steam from the hot copper vats. We were lost in the mesmerizing aromas and symphony of activities that were orchestrated to make this wonderful ricotta.

During our four-hour visit, we shared our farm story and exchanged lots of convivial personal history. Every member of their family was involved in cheesemaking, and after four generations, they seemed to have the process down to a science. A small storefront with refrigerated cases housed the cheese that was made that day for retail sale. After draining the ricotta, they promptly brought it to the front and began selling it at 9 a.m. Eagerly awaiting their fresh ricotta, the customers were ready. The ritual of gathering in person to purchase fresh food from local farms harkens back to a time that does not exist for most of us in the 21st century. The "farm-to-table" and "buy fresh, buy local" renaissance over the recent 25 to 30 years in the U.S. is a move back to community living and sustainable, local farm products our grandparents and great-grandparents knew well. It is a beautiful way to live and eat.

Our next cheese visit was scheduled for later in the afternoon before dark. We were surprised that Claudia and Lucia said little about the cheesemaker that we were visiting. We thought that was strange, but by 4 p.m., we were heading to a sheep dairy that included many agricultural initiatives on the their farm. The owners indicated to us they grew most of their own food   The only edibles they purchased were salt, sugar, and coffee. With darkness coming soon, we arrived just in time to see the end of milking. The father and son did all the milking chores by hand, which was not insignificant. Although we did not stand there and count, there appeared to be at least one hundred sheep in the herd. Most dairy herds that size would have at least an automated system, to collect and store the milk in refrigerated tanks, but not in this case. Father and son milked all the sheep by hand twice each day. They made cheese after the evening milking. Remember, by this time, it was dark outside, and they were just starting to make cheese. Unlike cows, sheep give about one-quarter to one-half gallon of milk per day. For 100 sheep, you would typically get between 25 and 50 gallons per day. This made sense to us, especially when you noticed that they made their cheese in a copper kettle over a wood fire which would hold enough sheep milk from the two milkings that day. In the U.S., cheese regulations would not permit the use of a wood fire in the cheese production area, but here in Sicily, it seemed like we were witnessing perfection with the combination of wood, fire and milk that represented the terroir of the farm.

They heated the milk and added natural rennet from the sheep's stomach to curdle the milk. After about an hour, the cheese was ready to be molded. Peppercorns were added to the cheese, making this one of the typical sheep milk tommes commonly sold as Pecorino Pepato. After making the cheese, they used a pulley system to move the copper kettle with the leftover whey back to the fire, where they started making ricotta. Nothing goes to waste in Sicily. The process of making ricotta is heating the whey and adding salt. A wine glass filled with salt caught our attention. We asked how much salt they were adding, and they looked at us amusingly and said a wine glass full. We thought we were going to receive our answer in ounces, cups, or some other metric like grams! That's just how they do it there. Traditions are important, recipes are not. Once the curds started to rise to the top, they skimmed off the ricotta and added it to molds to drain. Within the next 15 minutes, we were eating fresh ricotta at 7 p.m. at night. At this point, we assumed the day was over, but it would be premature to think that way. The owner's wife came out and removed the hot coals from the stove that were used to heat and cook the sheep milk cheese and ricotta. When we asked why, she informed us she was going to use these coals to cook the lamb that they harvested from the barnyard earlier in the day. The ease with which they dovetailed each of their activities provides a window into how Sicilians embrace conservation, sustainability, and using the surrounding land in the most efficient way that they can. These practices illustrated for us how and why they were nearly self-sufficient homesteaders. All the vegetables, fruits, meat, and dairy come from their land and labor.

The third cheesemaker visit was a big surprise, especially for Al. We were getting ready to leave the inn when Claudia informed us we were going to see the Italia family farm. Well, this was a shock because Italia is Al's grandfather's surname. There are only a few Italia families in Sicily, or even all of Italy. It is not a common family name. We subsequently used a website to search the Italia family name in Italy, Sicily, and Sardinia. There were only a few, but enough to convince us that Al's grandfather's surname was the real deal. Signor Italia looked a lot like Al's grandfather, and even resembled Al to some degree. When we explained that Al's grandfather was born in Sicily, the Italia's smiled with curiosity. They concurred that there are only a few Italia families in Sicily. Of course, he was not sure if they might be related, but he invited us to come back to spend more time researching genealogy and tracking down relatives who might remember or know of Al's grandfather.

While we were dating in the 1990s, Catherine first learned that Al's grandfather's last name was Italia. She was somewhat amused and said that after spending several years living in Italy, she had never come across a family name like Italia. She was living in Florence, but she had the opportunity to travel throughout Italy. Catherine theorized that, most likely, upon arriving at Ellis Island in New York, without speaking English, Al's grandfather might have been given this surname in America. Al never really thought much about Catherine's theory after that, so this turned out to be a big surprise and an emotional coming home for him that was not in the original travel plans.

Like many grandchildren of immigrants, we often wondered about our ancestors and the sacrifices they made to come to the United States, hoping to make a better life for themselves and their children. Al remembers his grandfather talking about Sicily and the fact that grapefruits were so plentiful, they would fall on your head as you walked by. His grandfather had the requisite Sicilian-like fig tree in his backyard that was diligently wrapped in burlap each year before winter. Al remembers harvesting bowls of figs, but never really understood why this was so important to his grandparents. It is not a far stretch of imagination to wonder if that fig tree came from seeds that they brought with them on the ship that carried them across the Atlantic Ocean back in the early 1900s. As a child, Al remembers not being interested in eating figs, but preferred processed Fig Newton cookies from the supermarket.

It's hard to understand in one visit exactly how farm families manage to get by. Mr. Italia described it well when he said, "We have horses, cows, sheep, goats, boar, and chickens. We have everything but cash!" Well, that pretty much sums up farming. It is not the most lucrative venture if you are trying to grow your bank account. It is a fulfilling lifestyle, working the land, co-working with the animals to provide food and sustenance. Then, the farm products, whether it is meat, dairy, olive oil, wine, or something else, meet the needs of community members, who in turn support the farm.

The Italia farm tour included the barns, paddocks, cheese room, and yogurt production area. We happened to be there when the younger son was making mozzarella balls for the upcoming farmers market. He was wearing an immaculate white shirt with a stylish haircut. His fit biceps and suntanned upper body were notable. His shy smile was hard to miss. Although the cheese and yogurt were good, we think there were other reasons this young man was a perfect fit for market sales. They say he sold out of dairy almost every week.

We cannot really explain the emotional impact of our Sicily visit except to say it felt like we had come home. It was not important to know with certainty if Al was related to this Italia family. What an inspiration, that after 60 years, here Al was meeting the next generation of farmers in his family's ancestral homeland. We realized that we were doing work in Pennsylvania, very similar to what the Italia family in Sicily was doing now. It was a full circle moment that left us humbled, inspired, and appreciative of what our ancestors did about a century earlier. We went to Sicily thinking about learning how to make better cheese, but what we left with was something more important. Despite the trials and tribulations of agriculture, being part of a culture valuing food and community brings people together. It inspired us to seek a new vein for understanding our neighbors, our friends, our country, and our world.

CHAPTER 13:

FARMERS EMERITI–
OUR NEXT CHAPTER

Time flies when you are having fun. We are so gratified and fulfilled by the 21 years (2001-2021) we enjoyed at Yellow Springs Farm. We have zero regrets about the journey, and similarly no second thoughts about ending our tenure farming in Chester Springs, Pennsylvania, in 2021.

People ask if the Covid pandemic prompted our decision. Surely it was a factor, but only one of many influences and realizations. When Catherine was working in the plant nursery in May 2021, she was wearing support braces on her left wrist and right elbow. The diagnosis for both painful joints was long-term, chronic overuse. Her lower back wrap had a longer history in falls from horses and subsequent surgery. Finally, the knee brace was associated with a deep groundhog tunnel opening that was disguised with long grasses, making it easy to overlook while scouting for ripe berries. The farm chapter has a mundane orthopedic history all its own for both of us. The sum of the days and years can simply be summarized as mileage—miles of life lived. Since we relocated to Maryland's Eastern Shore in 2022, a local metaphor of the watermen is apropos: "Ships are safer in the harbor, but ships are not made to stay at dock." Life is for living—dangers, risks, and tears are all part of the journey. Happiness is not a destination, but instead it is what happens along the way, with glimmers of joy and miracles of moments.

When we seized the "carpe diem moment" and began the Yellow Springs Farm adventure, we promised one another there would be a priority for a "next chapter" beyond the farm. We were resolute to farm while we were able to do so with open minds and wholehearted dedication. But we were equally honest to assess when it was time to move on before becoming overwhelmed by the physical work and the mentally taxing responsibilities of land management and business planning. Our patience was waning for daily morning rituals of waking up to see which projects had jumped from important to urgent. Checking texts before coffee, and often before sunrise, is not something we miss since leaving the farm behind.

Moving on to whatever this next chapter might become was a process. We did not wake up one day and say, "It's time. " We joked repeatedly over the years that we both did not want to quit the farm on the same day. We always supported each other when one of us was not feeling good about farm life, and needed private time to rebalance. The yearly rituals were things that we looked forward to and embraced. Even though January and February were easier task months, there was still much preparation for kidding and plant nursery season. Winter was for website updates, marketing plans and tax preparation. Goat kidding season in March and April kicked things off with sleepless nights and 24/7 chaos of unexpected births, late season snowstorms, and relentless bottle feeding schedules for the kids that were not nursing well. Even so, baby goats were always a joy with their floppy ears and sweet eyes. Spring birthing brought a flush of milk from the goat moms munching on new grass growing in the pasture. Making the first cheeses from this early spring milk was very exciting. Hearing the sounds of the vacuum pumps, the clicking of the pulsators mimicking the sucking action of goat babies on their mom's teats, filtering the milk, and making the first batch of cheese; it all seemed ethereal. There was something meditative about seeing and hearing the whey slowly dripping from the curd bags into the stainless steel drain trolley. Nothing can replace the singular herbal, vegetal, sweet aromas of fresh cheese.

April marked the nursery opening, and the uncovering of plants that overwintered while protected with white, felt-like frost blankets. We loved seeing new plant growth peeking through the soil. It was refreshing and cleansing to sweep away fallen leaves, right the turned over pots, and cut back last year's dried foliage. It was like opening day at the ballpark—everything was beautifully arranged –it seemed to ooze with potential and augur high expectations for the new season. As the weather teased us with a few warm days, we would receive our first nursery and landscape calls even weeks ahead of the last frost date. Everyone was eager to start planting. It was time to break out of our winter doldrums and witness the rebirth of nature. Spring and fall Open Farm days, the beginning and ending of our cheese CSA in May and November, farmers markets, and shipping holiday season orders galore were all a part of our yearly rituals. We never planned on any of this happening, but it did. The business growth was organic—no pun intended-- and we embraced the seemingly random path with expectations that these adventures would be fun.

Traditional farm families typically had children or other younger family members involved in activities to inspire generational transfer of the farm. For us, this was not an option. There were no children or nieces and nephews to take the reins of Yellow Springs Farm. In fact, Yellow Springs Farm itself, in hindsight, seems like the child we raised in our marriage. When people asked us whether we had children, we would say, "Well, not technically, but yes, if you count the hundreds of goat kids birthed during our tenure at the farm." And although we did not have a traditional farm family, we thought of the farm helpers that came through over 20+ years as our farm family. They impacted our lives, and we hope we made positive marks in their lives, even if in small ways. This potent "people connection" was unexpected, and it crept in the business plan little by little. It yielded a human crop, of sorts. Many of our farm helpers came as high school and college students looking to earn a few dollars and learn about agriculture, goats, and cheesemaking.

Others were adults who came during life transitions, testing the waters to see if agriculture was a career path worth considering, given their interest in farming and local food production. Most moved on to other non-farm endeavors that covered a range of careers, including engineering, medicine, public policy, landscape design, and government service. A select few, however, continued down the agricultural and sustainable food systems path, either directly or indirectly. It is inspiring to see them prosper locally, but also far beyond in Colorado, Minnesota, and North Carolina.

When we look back at making the decision to sell the farm, it was the little things that added up to eventually initiate this pivotal change of life direction. Everything that we enjoyed about the farm was still there--the community of supporters and friends, the land, the goats, and plant nursery. It was just that everything was getting a little bit harder. Producing our favorite cheeses became more of a chore than a pleasure. Open Farm Days were always a positive for us, since we received so much joyful interaction and affirmation, but the preparation time and clean-up seemed to be more arduous than it was the year before. Farm maintenance items that were once routine seemed to get a little larger and more expensive each year. We always pledged that when the positives of staying on the farm were outweighed by the negatives; it would be the right time to move on.

The time came in the middle of 2021 when we signed a realtor's contract to put our farm on the market. It was not just selling our property, but we were selling our home, and the place we created as the product our 21 year marriage. On quiet evenings, as we sat on rockers overlooking the pastures, we wondered if we were selling part of our souls. Preparing to show the property to potential buyers was a tremendous undertaking given we had accumulated all kinds of stuff. It was not just farm implements, cheesemaking equipment and nursery supplies, but also tons of baling twine, too many buckets, boots no one had worn, cardboard boxes both empty and full, papers, glass jars, platters, mismatched furniture, and countless more items we kept in case we might need them someday. Downsizing took months of second hand sales, giveaways, and Green Drop donations, but eventually we were ready to go.

We were fortunate to quickly find a property buyer, but unsuccessful in finding someone who wanted to continue to operate a goat dairy and /or a plant nursery. The new farm owners have no interest in continuing our farm business. They will simply enjoy the beautiful place. For them, this is likely a wise choice. Nonetheless, it is sad to see our cheese recipes idle, and the extra labels and marketing materials no longer used. Nothing is forever, so we hold a smidgen of hope that eventually, perhaps when we least expect it, we might meet an aspiring cheesemaker who would like to build on what we started.

The real estate sale contract sealed our intention, but left us with the urgent need to find a place to live for two adults and an 80-pound dog. The friends and relationships that we built in Chester County over several decades were irreplaceable. Our first thought was to downsize and find a home nearby or within 25 miles of Yellow Springs Farm. We were still at the height of the Covid crisis, and homes were in short supply. Two months of looking at houses in Chester

County did not make us a match. We placed bids on several houses, but each time we were outbid by other prospective buyers. Houses were selling for cash over listing price--no inspections, and fast closings. Chester County was not going to be a reasonable option in the short-term. Renting a house was a thought, but the idea of putting two moving trucks full of belongings in a storage unit did not sound financially appealing or practical.

We needed another plan. We did not at first have intentions to move to the Eastern Shore of Maryland, but given the difficulties of finding a house in Chester County, Maryland became an attractive alternative. We love the water, history and the small towns of St. Michaels, Easton and Oxford, in Talbot County Maryland. The area has culturally enriching art, music, and food scenes. It is home to many non-profit environmental conservation organizations. Its public policy embraces farming, land preservation and efforts to save the Bay on both sides of the political aisle. It seemed like a good fit for our value system. Over two decades, when we craved a quick get-away from the farm, we chose the Eastern Shore of Maryland. St. Michaels was the destination for our first "away date" before we were married, so it holds a special place in our hearts.

The property closing was scheduled for late December. Thanksgiving was approaching, and we still had not found a house to buy in Pennsylvania or Maryland. We talked about living in Italy for six or 12 months, but visa restrictions, and bringing a large dog overseas made that idea quickly evaporate. During the fall, we received emails from our Maryland realtor about homes that came on the market. The Eastern

Shore real estate market was similar to what we were experiencing in Chester County, PA - demand exceeded supply. Some listings looked attractive, but frequently we could not get to see them in person before the following weekend, and by then the houses were already sold. Winding down a cheese business, plant nursery and packing up a 19th century farmhouse and barn was all consuming, so we were bound to miss some market opportunities.

Our original aversion to renting was now negotiable, so we planned to see several rentals near St. Michaels on the Friday after Thanksgiving. Winter was off-season here, so there was plentiful options. If we had to wait for a desirable house to come on the market, why not plan an extended vacation on the Eastern Shore of Maryland?

As we were on our way to Friday showings, we received a call from our realtor. One of the houses that we liked a month or so earlier just came back on the market. The prior deal did not go forward for some reason. We jumped at the chance to see the house, and made it our first stop. This house would not stay on the market for long and we would have to make a quick decision. Promptly at 9am, we met the realtor, entered the house and smelled the soothing lavender aroma strategically used in the home staging. It was an easy choice. It was the house we dreamed of over the years as we visited the Eastern Shore, but never thought we would own. The floor to ceiling windows filled the house with daylight. The great room and screened porch faced west, assuring us countless sunset views over Harris Creek. Catherine's artistic spirit was overwhelmed with joy as she imagined the possibility to paint here. The home had large rooms, and

updated baths and kitchen which we definitely preferred after living in an 1850's farmhouse for over two decades. It
took a few weeks of inspections and paperwork in order to
finalize the contract. Then, we walked to the post office to
rent a mailbox anticipating our move.

December was all about packing. The cleaning and
preparation to move seemed endless. Our realtor, Linda, still
a friend, was a life saver. She was there shoulder-to-shoulder with us every step of the way. We were exposed to Covid
while visiting Catherine's elderly parents for Christmas just
days before the move. This was before home tests were readily available, so we searched for drive-up, quick turnaround
Covid tests. It was a weekend, so we paid a premium. The
test results came back negative; what a relief that we would
not get sick, or be contagious to the movers coming the following day. Two large trucks were loaded with our belongings. We filled both of our cars with everything that could
not get on the trucks. Roscoe, our dog had a tight fit in the
back seat, but did not seem to mind the clutter once he settled in. As the moving trucks pulled away, we looked back
one last time. It was difficult and poignant. A sense of relief
cascaded over heavier feelings of exhaustion, mourning,
apprehension and anxiety in these last moments at Yellow
Springs Farm. Our final act of ownership was to together
remove the farm sign near the road, next to the mailbox, as
we were leaving.

The farm prepared for a new chapter, too, like a capable
young adult ready to head in a new direction. Because we
donated a conservation easement to a land trust to protect
the farm's landscape in perpetuity, and further enrolled the

parcels in the state's Agricultural Security areas, everything was in place for passing the baton of stewardship to new owners. Children go to college or move to independent living after about 20 years. This similar time frame is significant for us in raising the farm together. We gave the farm roots and wings to fly to its new owners. Let's hope they love the land. As former caretakers and stewards of the farm's historical and agricultural legacy, we want to see the farm grow, change and fulfill dreams for future owners, but from a distance.

Catherine craved more time to paint with uninterrupted hours in the art studio. She now exhibits her work regionally throughout the year, and is gratified when collectors are inspired to purchase her oil paintings. Al was excited to go fly-fishing whenever the weather was right, instead of only when farm chores allowed a flexible schedule, typically in winter. Travel dream lists were made and only metered by finite amounts of money and an unknown number of years ahead. This description of our next chapter sounds like retirement aspirations of almost any couple of a certain age, but we don't consider ourselves retirees.

When people ask us now if we retired after selling the farm, we gently respond with a description of our desire to build a new life chapter. Yellow Springs Farm was a lifestyle. It was a place that existed with a soul of its own long before we lived there. This 21-year connection to the land, people, plants, animals, aromas, and stone buildings on the Farm imprinted our hearts and minds much like a relationship with a treasured pet dog, cat, or horse. Even when the pet passes, it forever changes its people. The farm has had that larger-than-life place in our life so there can be no next chapter without the farm, but more specifically a new relationship to enjoy the ephemeral aspects of the farm and to nurture the connections to people and places that persist as treasured holdings long after the real estate sold. Although we no longer own a farm, we continue to operate the business entity "Yellow Springs Farm LLC." The name and the website URL stay with us. Our business now is largely a service business. We don't work every day, and we have plenty of autonomy in making work schedules. We offer landscape design and consulting, business consulting, and help producers develop and market value-added agricultural products. We are event speakers, webinar instructors, and with this book, we are authors.

If a job is just a job, separation is possible after resigning and passing a transition period measured in weeks. This transition will take longer. Al still has flashbacks about the 4 a.m. alarm ringing to get the cheese turned in the brine, or the milk parlor set up before sunrise. Catherine freezes when the phone rings, echoing anxieties about a cheese order that did not arrive in time, a fence broken with goats on the run, a doe experiencing complications in labor, or a vehicle breakdown with a stranded farm employee. These are worries and responsibilities we volunteered for—no martyrdom—but we have no regrets now as *Farmers Emeriti.*

These days, a few houseplants and border gardens take the place of a plant nursery, pastures, and woodlands. A family of resident bluebirds is the closest thing to livestock seen in our new residential life. We named the matriarchal bluebird "Pina." She honors the space in our hearts for Rena, the first doe born on the farm that produced so many daughters in our herd. We kept promises to one another to assure there is a new chapter beyond the farm, but we will never be without the farm influences. Looking back on the first years of post-farm life, visitors to our new home in Maryland include former farm volunteers, ex-employees, and CSA members. The farm community of friends, colleagues, and customers really was a chosen family.

Three pivotal relationships have formed in Maryland since our move. Catherine was excited to become a customer at an organic farm - Cottingham Farm in Easton, Maryland. The greens, eggs, and pork are delicious, nutritious, and photogenic. The landscape enjoyed driving to Cottingham Farm is reminiscent of Chester County, with agricultural outbuildings, field rows sown, and tractors working the land. We quickly befriended Cleo and Allie, who own and run the business. Catherine even volunteered to staff the farmstand, but just occasionally. Commonalities overflowed, and there was quick empathy and shared values with these fellow producers of healthy, fresh food. As a foursome, we are all career-changers who share enough experience and eccentricity to write yet another book. We became part of Thanksgiving traditions at Cottingham Farm in 2022 and 2023. We are grateful to commiserate, celebrate, and collaborate with new farm friends.

Saturday mornings without farmers markets just don't seem right. We shop regularly at the outdoor, markets in Easton, St. Michaels, and occasionally Chestertown, Maryland. Catherine had lunch with Chef Jordan Lloyd, who was the St. Michaels Farmers Market board president in spring 2023. It was only a matter of time before she volunteered to join the market board. Seeing the market from the other side of the table is refreshing. Advocating for vendor opportunities, customer offerings, and community building comes naturally after being vendors for two decades. It's not an exaggeration to say that among the non-negotiable prerequisites for our new home were a first-floor bedroom, space for a nice art studio, and, of course, a good local farmers market.

We love to eat. We live to eat, rather than eat to live. Perhaps that is assumed given our Italian ancestry and dedication to artisanal cheesemaking. Without the farm, our food hobbies are now means to make new friends, enjoy "couples-time" in the kitchen, and seek opportunities to buy local farm products. It's fun to support small businesses and restaurants that purvey delectable goodies in convivial spaces. Former farmers like us make grateful patrons at all food-related establishments, but especially well-stocked cheese counters that reflect the proprietors' curious, if not adventuresome, palates. Our appreciation just grows grander for what it takes to bring goodness from field-to-fork, from farm-to-table, and now in Maryland from shore to sandwiches.

Our Maryland agriculture and aquaculture crops—all non-commercial—now include pollinator plants for local bird and insect habitat, and oysters quietly growing in cages off the pier in our rear yard. In a stroke of synchronicity, we live on Harris Creek which is a Maryland-designated oyster sanctuary. The water quality is exceptionally good thanks to abundant sub-aquatic vegetation, especially widgeon grass.

The oysters help keep the water clean, and they thrive in this brackish, tidal environment. Many of these same conditions make Harris Creek a great place for late summer crabbing. Our *aperitivo* hour was once always a wine and cheese event, but there are new expressions of terroir in oysters and crabs harvested steps from home. The Chesapeake Bay region is culturally different than Chester County, PA. The focus here is on the water, the coastline, sailing and motor boating. Neighbors are passionate about fishing, especially for rockfish, and making a living on the water. The cultural dynamic between the 'from here" and the "come here" populations colors conversations.

The interconnected people and community aspects of our farm life in Pennsylvania continue here in Maryland, both on a local and broader scale. The local food movement is thriving here but with similar producers 'challenges of not having enough capital, infrastructure, and regulatory know-how to effectively serve customers. On a larger scale, the Chesapeake Bay watershed extends from upstate New York to Norfolk, Virginia, encompassing 64,000 square miles. With continuous industrial and residential development over decades, there are now approximately 18 million people living in the Chesapeake Bay watershed. Rising sea level is real, and mitigation efforts are underway in several small towns on the Eastern Shore. Efforts to improve water quality by planting billions of oysters and subaquatic vegetation will contribute to maintaining and hopefully improving the health of the Bay, but in our limited time here, it is apparent that managing multiple state agendas and other governmental, public, and private institutions dedicated to rehabilitating the Bay will be challenging for decades to come. Local restoration of the 100,000 streams that feed the Chesapeake Bay and its tributaries throughout the watershed is evolving into more of a priority. Stopping pollution and sedimentation locally will increase the chances of success for downstream restoration efforts. Creating local food systems in our communities and continuing to fund and support local watershed system restoration efforts could result in significant improvements for the Bay.

The word "harvest" is important on the Eastern Shore, given that much of the land is dedicated to agricultural pursuits. As a couple starting a new life chapter in Chester County, we are ready to reap what we sowed. Over the course of 20-plus years, we turned our dreams into a beautiful place, lifestyle, and business. Now, we have no animals to feed, no employees to manage, and no orders to deliver. This is a time to refresh, renew and explore. Website revisions (www.yellowspringsfarm.com) chronicle our personal evolution. First, in 2022 after we sold the real estate, we changed the mailing address and updated the business timeline. 2023 website changes feature our agricultural consulting services, and in 2024 we highlight more of Catherine's paintings, and give a special shout out for our book. We tried to count how many website re-dos there have been since 2001. Let's just say, "Too numerous to count." Yellow Springs Farm LLC plans never followed a straight line or a rigid matrix. Then, as now, we are changing, morphing, and tweaking things. Our careers have not ended, but the farm property is behind us as we look forward.

When advising our landscape customers what to expect

from their newly planted gardens, we use the idea that in the first year, plants "sleep", in the second year they "creep" and in the third year they "leap." We faced unexpected health challenges, plus expected orthopedic maintenance items that slowed us from integrating and building new relationships in our first two Maryland years. Now in the "leap" phase, our third year, we started 2024 speaking at agriculture conferences in Pennsylvania and Maryland; teaching a webinar on how to commercialize value-added agricultural products, like cheese and jam, in collaboration with the Chesapeake Agriculture Innovation Center (CAIC); designing a few residential native plant gardens; and finally finishing and promoting this book. Our 2025 calendar is already filling up.

It is not our temperament to sit still and wait for life to come to us. It is easier to connect the dots and understand how things happened and fell into place when looking backwards with hindsight, but close to impossible to foresee events. Visits to the Eastern Shore for brief respites over the last 20 years somehow indirectly, perhaps subliminally, led us to the place we now call home. There is a reason we are here. Time will tell. We are unsure what the future holds, and that is okay. This next chapter of our life is a work in progress. A cup full of possibility is promising abundance.

www.ingramcontent.com/pod-product-compliance
Lightning Source LLC
Chambersburg PA
CBHW060543210326
41519CB00014B/3330